LUNAR ROVER

1971–1972 (Apollo 15–17; LRV1–3 & 1G Trainer)

'One of the finest-running little machines I've ever had the pleasure to drive.'

Gene Cernan, 14 December 1972. Amongst the last words ever spoken on the lunar surface.

© Christopher Riley, David Woods & Philip Dolling 2012

All rights reserved. No part of this publication may be reproduced or stored in a retrieval system or transmitted, in any form or by any means, electronic, mechanical, photocopying, recording or otherwise, without prior permission in writing from Haynes Publishing.

First published in 2012

A catalogue record for this book is available from the British Library

ISBN 978 0 85733 267 7

Library of Congress control no. 2012940365

Published by Haynes Publishing,
Sparkford, Yeovil,
Somerset BA22 7JJ, UK.
Tel: 01963 442030 Fax: 01963 440001
Int. tel: +44 1963 442030
Int. fax: +44 1963 440001
E-mail: sales@haynes.co.uk
Website: www.haynes.co.uk

Haynes North America Inc.
861 Lawrence Drive, Newbury Park,
California 91320, USA.

Printed in the USA by Odcombe Press LP,
1299 Bridgestone Parkway,
La Vergne, TN 37086.

Comment on Units

Apollo engineers worked in both imperial and metric units. This was partly due to strong European influence on the programme. Many of the diagrams in the following pages are in inches, and the Rover's weight on Earth was often quoted in pounds, whilst distances and speeds travelled by the Rover were calibrated in metric and recorded in kilometres and kilometres per hour.

NASA engineers were often surprisingly vague as to whether they meant pounds-mass, pounds-weight or pounds-force. Writing about the gravity fields of Earth and the Moon in telling this story of the LRV, we have adopted kilograms to imply a definite statement of mass, not weight. But elsewhere we switch to imperial units, reflecting the nature of Apollo era engineering practice. The table below is provided to help you grasp the figures quoted in either standard.

Distance		
feet	0.3048	metres
metres	3.281	feet
kilometres	0.6214	statute miles
statute miles	1.609	kilometres
nautical miles	1.852	kilometres
kilometres	0.54	nautical miles
nautical miles	1.1508	statute miles
statute miles	0.86898	nautical miles
Velocity		
feet/sec	0.3048	metres/sec
metres/sec	3.281	feet/sec
metres/sec	2.237	statute mph
statute mph	0.447	metres/sec
feet/sec	0.6818	statute mph
statute mph	1.4667	feet/sec
feet/sec	0.5925	nautical mph
nautical mph	1.6878	feet/sec
statute mph	1.609	kmh
kmh	0.6214	statute mph
nautical mph	1.852	kmh
kmh	0.54	nautical mph
statute miles/sec	1.609	km/sec
km/sec	0.6214	statute miles/sec
Volume (capacity)		
US gallons	3.785	litres
litres	0.2642	US gallons
US gallons	0.833	imperial gallons
imperial gallons	1.201	US gallons
cubic feet	0.02832	cubic metres
Mass		
pounds	0.4536	kilograms
kilograms	2.205	pounds
Pressure		
pounds/sq inch	70.31	grams/sq cm
grams/sq cm	0.0142	pounds/sq inch
Force		
pounds force	4.4482	newtons

LUNAR ROVER

1971–1972 (Apollo 15–17; LRV1–3 & 1G Trainer)

Owners' Workshop Manual

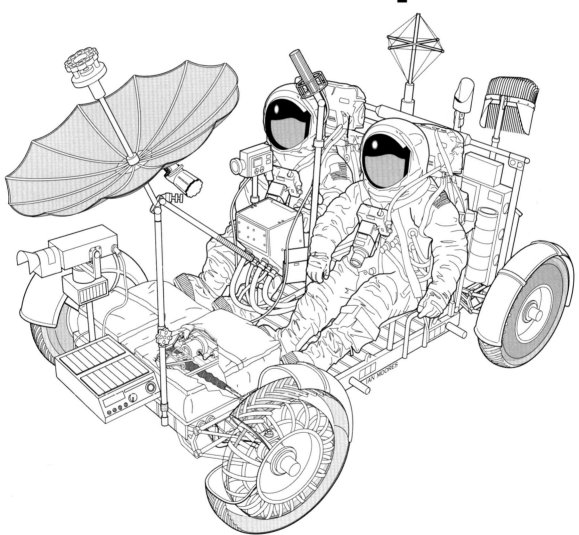

An insight into the technology, history, development and role of NASA's unique Apollo Lunar Roving Vehicle

Christopher Riley, David Woods & Philip Dolling

Foreword by David R. Scott, Commander of the Apollo 15 mission

Contents

6	Foreword

8	Introduction

First dreams of driving on the Moon	13
Project Horizon	16
A robotic vehicle for the Moon	18
The birth of NASA's manned lunar mobility programme	25
MOLAB	26
Scaling back	27
Persuading NASA	31
A new competition	33
The final entries	35
NASA's choice	37
The poisoned chalice	38

40	Structure

Eight engineering systems	42
Weight and strength	44
Failure is not an option	47
A new method of deployment	49
A change of plan	49

52	Mobility

Where no wheels had rolled before	54
Wheels	55
Motors	59
Motor control	60
Transmission	62
Steering	64
Suspension	68
Brakes	72

74	Electrical and thermal control

| Electrical power | 76 |
| Thermal control | 83 |

OPPOSITE The prime crew of Apollo 15 pose for a pre-flight photo on the 1G Rover, in front of a Lunar Module mock-up From left to right they are James Irwin (LMP), David Scott (Commander) and Al Worden (CMP). Worden has his left hand on the Apollo subsatellite, which would be ejected from the Command and Service Module in lunar orbit. *(NASA)*

92	Navigation & communication

Navigation	94
Communications	102
Television	105

112	Crew station & Control and Display Console

Getting it right first time	114
Crew station	114
Control and Display Console	122

128	Stowage and deployment

Stowage	130
Deployment	132
Rover deployment and start-up instructions	134
To the Moon	151

152	Wheels on the Moon

Deployment on the Moon	154
Starting the car	156
Navigation system performance	157
Off-road driving	157
Hill climbs	159
Hillside stability	160
Going downhill	161
Speed demons	162
Suspension and ride	162
The T-handle	165
Blinded by the light	165
Breakdown and recovery	167
Keeping cool	173
Juggling batteries	175
The blessings of the Rover	175
Interplanetary outside broadcast	176
At the VIP site	181
Rest in peace	184

186	Epilogue

190	Glossary

192	Acknowledgements

LUNAR ROVER MANUAL

Foreword

"LRV-1" – Apollo Lunar Roving Vehicle No. 1.
Crew: Apollo 15 (David Scott and Jim Irwin)
First outing of LRV-1: July 31, 1971; duration 5½ hours;.
Second outing of LRV-1: August 1, 1971; duration 6½ hours.
Third outing of LRV-1: August 2, 1971; duration 4½ hours.
Total distance travelled: 27.9km.
Total payload returned (rocks and soil): 77.3kg.
Disposition: remains parked near Hadley Base, Hadley Apennine region of the Moon.
Availability: battery replacement required (however, prior to operating LRV-1, an *Owner's Workshop Manual* should be available and studied).

OPPOSITE Astronaut Jim Irwin took this shot of Apollo 15 Commander, David Scott, in the driving seat of LRV-1 at the start of their second day of exploration at Hadley. *(Jim Irwin/NASA)*

In December 1969, I was assigned the Commander of the planned fifth lunar-landing mission, Apollo 15, an advanced "H"-type mission, similar to Apollo's 12, 13, and 14 – two exploration traverses, each walking approximately 1km from the Lunar Module (LM). Having been the back-up Commander on Apollo 12 (November 1969), I looked forward to the new backpack and our preliminary landing site at Davy Rille. Jim Irwin and I would be able to double the walking time and distance from the Lunar Module and explore one of the most important features on the Moon – a large rille, or canyon. But we would be limited by what we could carry individually – tools outbound and samples returned.

Four months later, in April 1970, as our training was reaching high gear, the most dramatic and hazardous halt to the programme occurred: the near-loss of Apollo 13. After the spectacular rescue of the crew, the questions began to come: should the programme continue? Should we take any more risks? Should we terminate Apollo and be happy with our previous successes? Or should we go on? At that time, the design of a Rover had commenced, but the programme was in serious trouble. It was behind schedule, over budget, and not satisfying its basic requirements. Termination was being seriously considered.

But then, two months later, in August 1970, NASA made one of its boldest decisions. In the face of the near disaster of Apollo 13, dwindling public support, and a rapidly declining budget, NASA decided to skip the final "H"-type mission, press on with upgrading the total "system" (hardware, software, science, and operations) to the "J" configuration, and launch *three* full-up "J" missions to the most significant scientific sites on the Moon that contained diverse geological formations and materials within local areas. However, for human explorers to reach these, multiple missions had to be flown, or some form of mobility was necessary. The Rover emerged as the most practical solution to this challenge. Therefore, this upgrade from "H" to "J" included a full commitment to the LRV.

One month later, on 2 September 1970, Apollo 15 was redesignated a "J" mission, including the first "flight" of the LRV. The vistas for scientific exploration on the Moon opened wide, and Apollo 15 was assigned a site that included unique geological features that could be explored with a Rover. The engineering challenge was ominous, but within 11 months, we were driving LRV-1 at the Hadley Apennine site, one of the most rewarding sites of the entire lunar-exploration programme.

Now, after many years, the evolution of this engineering challenge and the resulting capabilities of the Rover are described in crisp detail in this remarkable *Owner's Workshop Manual*, which will become a classical reference for explorers, historians, researchers, engineers, scholars, and students for decades to come. And it will most likely become the basis for both the design and the operations of a human planetary-exploration wheeled vehicle, a device that will be used, in one form or another, for all future lunar, and planetary, explorations. Both the scientific and engineering communities appreciate the careful and thorough efforts of Philip Dolling, David Woods and Christopher Riley in compiling this valuable document.

David R. Scott
Los Angeles
June 26, 2012

"From the nature of things, astronautics can have no final goal; there will always be new frontiers beyond the furthest range of man's exploration."
Arthur C. Clarke, 1954

Introduction

This is the story of humanity's first drive on another world, just ten short years after the first human being orbited the Earth and only two years after men first walked on the Moon.

OPPOSITE LEFT Apollo 17 commander Gene Cernan's spectacular panorama of Station 5, across the Camelot Crater rim boulder field towards Wessex cleft. Harrison Schmitt can be seen running back to LRV-3 carrying the scoop.
(Gene Cernan/NASA/David Woods)

RIGHT **Wernher von Braun**, looking out from a tenth floor window of building 4200 at the Marshall Space Flight Center. *(NASA)*

BELOW Apollo 17 mission commander Gene Cernan makes a short checkout of LRV-3 at the start of EVA-1 in the Taurus-Littrow valley. This 'stripped down' view of the rover shows it prior to load-up. The mountain in the background to the right is the South Massif. *(Harrison Schmitt/NASA)*

Engineered and constructed in a record time of just 17 months, the shortest development and manufacture of any major piece of Apollo equipment, three Lunar Roving Vehicles (LRV) would eventually carry six men a total of almost 90km across the lunar surface. Reducing the exertion required for a moonwalk, these vehicles would double both the amount of time the astronauts could spend exploring and the total weight of lunar samples that could be collected; thereby maximising the scientific value of the Apollo programme and paying back the Rover programme's entire $38 million development costs many times over.

At the time of its development, when battery powered electric cars were thought of as an oddity rather than an ambition, the concept of an electric lunar vehicle was as visionary as the Apollo programme itself. As Gene Cernan, the last man to drive a Lunar Rover on the Moon put it, "the president had plucked a decade out of the 21st century and inserted it into the 1960s and '70s". And nothing epitomised such forward thinking more than this visionary lunar transportation system.

A wheeled vehicle was not strictly necessary for a quick trip to the Moon. Without them Apollo would still have been an amazing chapter in human history. But the Lunar Rover embodied a deeper commitment by NASA to human space exploration. Complete with manuals packed with data on performance, tips for operation, and carefully conceived procedures for maintenance they were a vision for a long-term presence on a new world.

For one man, this capability of driving across the Moon was a logical extension to landing men there. As early as 1963, barely one year into the Gemini programme, German-born engineer and father of the Saturn V rocket, Wernher von Braun, firmly told an astronaut rookie called Gene Cernan that one day he'd be driving on the Moon. With human spaceflight still in its infancy this was a bold prediction which even Cernan – a gung-ho Navy aviator – didn't believe would happen. But sure enough, within a decade, that's exactly what Cernan was doing; pushing the third Lunar Roving Vehicle to record speed through the spectacular Taurus-Littrow valley.

Without such a strong conviction in the need for them, these extra-terrestrial vehicles might never have been built; for their inclusion in the Apollo programme was constantly in jeopardy.

After years of uncertainty and hesitant commitment to their development, NASA finally commissioned a Rover for Apollo in the middle of 1969; just weeks before any human even set foot on the Moon.

The vehicle would have to be extremely light, weighing less than a sit-on lawnmower, yet be able to carry twice its own weight in men, life support equipment, tools, samples, and

RIGHT **The crew of Apollo 13 step out on to the deck of carrier USS *Iwo Jima* after their brush with death.** *(NASA)*

science kit. In comparison, the average family car can carry only one-third to one-half of its own weight. The Rover would need to fold into a cramped compartment on the side of the Lunar Module that would carry it to the Moon's surface. Its deployment mechanism would have to be operated quickly and effortlessly by space-suited astronauts on a very tight schedule.

It would have to withstand the rigours of launch and transportation across 400,000km of space, exposed to a hard vacuum and extreme temperatures. At its destination, still in a vacuum, it would be bathed in fine, highly abrasive and tenacious dust. Its suspension system and tyres would have to operate on a world where the gravity is only one-sixth as strong as Earth's and whose intensely cratered surface would tax the most robust of all-terrain vehicles.

All this would have to be designed, prototyped and manufactured within a tight 17-month window. It was only the decade of lunar transportation research which von Braun had steered America into reluctantly carrying out that enabled such an undertaking to be embarked upon with any chance of success.

But even as the Rover was taking shape, just a year before delivery to the Kennedy Space Center, momentous events unfolding 320,000km away on board Apollo 13 threatened to halt the Rover's production, and called the entire future of the Apollo programme into question.

Three nail-biting days after that accident, with the crew safely back on Earth, curtailing Apollo without loss of life in space looked tempting. But instead NASA reaffirmed their commitment, not only to go back to the Moon four more times, but to make the last three flights J-class exploration missions that would carry Rovers with them.

Against the odds, the first vehicle was delivered to the Cape in time for the flight of Apollo 15 in the summer of 1971. It allowed Dave Scott and Jim Irwin to travel four times further from their Lunar Module than crews of previous missions had managed. Riding between field locations on LRV-1, Scott and Irwin were able to reach three different geological provinces during the precious eighteen hours available to them outside their lander, resulting, as hoped, in the discovery of the oldest rock samples collected at that point in the programme.

With each new Rover-supported Apollo mission, records were broken. Charlie Duke and John Young on their Apollo 16 mission clinched a new lunar-land-speed record in April 1972,

BELOW **Arrival of LRV-1 at the Kennedy Space Center during the spring of 1971.** *(NASA)*

TOP **A panorama showing LRV-1 at Station 6a, with commander Dave Scott using the sighting scope on the high gain antenna. The TV camera is pointed down toward the right, front wheel. Some of Jim Irwin's footprints can be seen near the bottom of the picture.**
(Jim Irwin/NASA/David Woods)

ABOVE **Charlie Duke's Station 9 pan, showing John Young with the dust brush at the front of LRV-2, probably just before he moves around to dust the TV lens. He has the hammer in the pocket on his right shin. Stone Mountain is in the distance.** *(Charlie Duke/NASA/David Woods)*

BELOW **120 hours, 48 minutes and 56 seconds – taken by Harrison Schmitt. Gene Cernan is getting the jack-and-treadle tool from LRV-3 to help them extract the deep core from the ground.** *(Harrison Schmitt/NASA/David Woods)*

hitting 17kmh driving downhill. This record was broken by Apollo 17 in December, when Gene Cernan reached 18kmh on another downhill run. His record remains unchallenged!

Apollo 17 also set a new single-sortie distance record when LRV-3 covered 20.2km in one day. This last rover mission traversed a total of 34.8km; a cumulative extra-terrestrial driving record that would last barely four months until it was surpassed in April 1973 by an unmanned Soviet rover called Lunokhod 2 (see photograph on page 76).

On all three of the Lunar Rover's operational drives, totalling 89.3km, they never let down their operators. Other than the occasional temporary failures of their front-wheel steering systems which were remedied by reverting to rear-wheel steering, and Cernan's breaking of a fender which was repaired on the Moon with some duct tape and a map, nothing went wrong with them.

As products of the 1960s these vehicles were built for durability not for disposability. Forty years on, those who designed and drove them on the Moon are quick to remind you, with a twinkle in their eyes, that should anyone return to the places they were finally parked, taking with them a new set of batteries and electronic components, the vehicles could

ABOVE **Commander Gene Cernan stands in front of a field of boulders SSE of Station 2, below the South Massif slope, where LRV-3 is parked.** *(Harrison Schmitt/NASA/David Woods)*

still be driven away. Perhaps one day they will be visited again, perhaps even retrieved for posterity and displayed in museums back on Earth – or even on the Moon itself – as monuments to the audacity of the engineers who dreamed of driving on other worlds in the mid-twentieth century and then made it happen.

Gene Cernan parked the last Lunar Rover for the final time in the Taurus-Littrow valley on the morning of 14 December 1972. His last comments on stepping off the vehicle left those who'd designed and built them brimming with pride. "One of the finest little machines I've ever had the pleasure to drive," he declared. They were amongst the last words ever spoken on the lunar surface.

First dreams of driving on the Moon

As Arthur C. Clarke, the British scientist and science fiction writer, noted in 1954, "probably no achievement of the human mind has been so well documented before the event as has the conquest of space". Such detailed planning and forethought of what would eventually come to pass is exemplified by the dreams of driving across other worlds.

Spurred on by an insatiable curiosity and vivid imagination, our first designs for a lunar rover emerged around the time that the motorcar started to be mass-produced at the beginning of the 20th century.

One such design was recorded in a 1901 science fiction novel called *Srebyym Globie* (On the Silver Globe) by the Polish writer Jerszy Zulawski. His rover had a pressurised cabin and was electrically powered with a top speed of 10kmh. Its tractor-like wheels could be removed and replaced by legs or "claws" for climbing hills or traversing particularly rough surfaces. The vehicle could even double up as a boat. In a curiously prophetic storyline, one of the characters in Zulawski's tale of lunar exploration,

LEFT **Jerszy Zulawski's Rover concept from his 1901 science fiction novel** *Srebyym Globie.* *(BIS)*

RIGHT Hugo Gernsback's 1915 rover, described in his book *Baron von Munchausen's Scientific Adventures*. *(Courtesy Rob Godwin)*

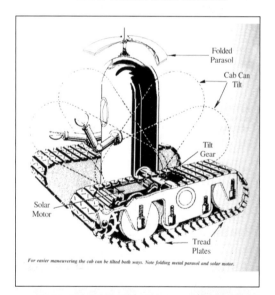

RIGHT Hugo Gernsback's Homomobile Lunar Rover concept from the 1960s. *(BIS)*

RIGHT Konstantin Tsiolkovski's 1918 lunar rover concept. *(BIS)*

FAR RIGHT Homer Eon Flint's 1923 rover from his novel *Out of the Moon*. *(BIS)*

who withdraws from the mission just before launch, was called Braun.

Baron von Munchausen's Scientific Adventures, written and published by the Luxembourgian-American author Hugo Gernsback in 1915, also documented a lunar exploration vehicle. The father of science fiction, as Gernsback was sometimes known, dreamed up a strange pressurised spherical steel chamber 18m in diameter, which would travel the Moon's surface on a track wrapped around its circumference. Gernsback's roving vehicle also housed the crew during their journey to and from the Moon. Almost 50 years later he would come up with another lunar rover design, which he dubbed the "Homomobile". It came complete with tank treads for movement, mechanical arms and a folding parasol for protection against the Sun!

The Russian pioneer of astronautics, Konstantin Tsiolkovski, also described a lunar rover in his early novels. His 1918 story *Vne Zemli* (Outside the Earth) records a rover that doubled up as a lunar lander. Tsiolkovski's pressurised two-seater vehicle had electrically powered wheels and small rockets for boosting it over crevasses. It also carried a larger rocket to enable it to lift off and return to Earth.

Some less serious designs for a lunar rover emerged in the science fiction writings of the 1920s and 1930s. A notable example is the rather ungainly bi-pedal "walking rover" concept from Homer Eon Flint's 1923 novel *Out of the Moon*.

By the 1950s some concepts were getting closer to a realistic rover that might actually be carried to the Moon to aid our exploration. In his 1951 book *The Exploration of Space*, Arthur C.

Clarke proposed a lunar surface transportation system equipped with large pneumatic balloon tyres. Clarke's prediction was that such vehicles would be powered by electricity from either storage batteries or gas turbines driven by reacting rocket fuels.

Wernher von Braun's 1953 paper on the subject envisaged a tracked lunar rover driven by hydrogen peroxide powered steam turbines, similar to those with which he had experimented while developing the V2 missile during World War II. Around this time von Braun, writing with astronomer Fred Whipple and science fiction writer Willy Ley, published an article in *Collier's* magazine describing detailed plans for spending six weeks on the Moon with three 10-ton articulated trucks carrying supplies.

The following year, in his book of collected drawings *Exploration of the Moon*, the self-taught English engineer and draughtsman Ralph A. Smith published a number of his own pictures of lunar exploration, which included Moon rovers. Unsure of the exact nature of the lunar surface, he'd conceived the vehicle with very large wheels, both to negotiate rugged and broken areas and also to reduce the loading imposed on the potentially weak dust-covered lunar surface.

A series of far less practical designs also emerged in this mid-20th-century period. In *Popular Science Monthly*, Frank Tinsley described an autonomous unicycle vehicle with a spherical body inside its rolling rim, and composed entirely of light, inflated fabric parts that would be easily disassembled and deflated for storage.

Another inflatable proposal, designed by the Scully-Anthony Corporation of Chicago, suggested that lunar explorers would inflate their barbell-shaped "Moon Sac" and then roll around the surface, breathing the air inside it.

But perhaps the most outlandish concept of this time came from German space pioneer Hermann Oberth. Published in *Man in Space* in 1954, his somewhat unbalanced design required gyroscopic stabilisation. It consisted of a spherical pressurised cabin 5m in diameter, fixed to the top of a column 5m above its caterpillar tracks. An elaborate mirror system perched on top of the cabin harnessed solar power, although the vehicle's main power

LEFT R.A. Smith's 1954 rover, from his collection of drawings *Exploration of the Moon*. *(BIS)*

CENTRE Frank Tinsley, writing in *Popular Science Monthly,* described an autonomous unicycle vehicle with a spherical body mounted inside its rolling rim, and composed entirely of light, inflated fabric parts easily disassembled and deflated for storage. *(BIS)*

BELOW Cutaway drawing of a lunar vehicle proposed by Antony-Scully showing how a two-man team would provide mobile power through a treadmill-type foot operation. Hand ropes would help them to maintain balance. To turn the "Moon Sac", one man would keep walking while his partner remained stationary. *(BIS)*

RIGHT Hermann Oberth's outlandish 1954 rover concept published in *Man in Space*. (Courtesy Rob Godwin)

RIGHT R.A. Smith's small, unpressurised rover idea from the early 1950s, resembling the famous World War II jeep. (BIS)

BELOW A lunar outpost concept devised for Project Horizon as late as 1965. (US Army)

came from a hydrogen peroxide engine. Not a man to be overly concerned with engineering practicalities, Oberth's unlikely lunar rover weighed in at 1,654kg on the Moon and could travel at speeds of up to 150kmh!

Far closer to a concept for the Lunar Roving Vehicle that would one day become a reality, was an early 1950s drawing by Ralph A. Smith for a much smaller open, unpressurised design which resembled the famous World War II jeep. Being English, Smith's rover was a right-hand drive, whilst the eventual Apollo LRV would have centre-mounted controls to allow either astronaut to drive. Smith died in February 1959 and would never know how close he came to envisioning the future of manned lunar exploration.

Project Horizon

In early 1959 the US Army Ordnance Missile Command (OMC) in Huntsville, Alabama undertook the first serious study of a lunar transportation system as part of the Army Ballistic Missile Agency's (ABMA) plans for a manned Moon base.

As head of the ABMA, Wernher von Braun assigned Heinz-Hermann Koelle to head the operation, which was code named Project Horizon. Plans envisaged a lunar outpost capable of housing 12 men becoming operational by December 1966 at a cost of $6 billion; although this may seem like a lot of money, especially for that time, it was less than 2 per cent of the annual budget of the Department of Defense.

Koelle assigned the task of developing a manned lunar rover, for use in conjunction with the Moon base, to the Army's Transportation Corps, which in turn approached General Motors (GM) for advice.

A young motor vehicle engineer called Sam Romano, heading the Special Vehicle Development team at the company's new Defense Research Labs in Detroit, recalls their request. "The Army came to us and said, 'Hey, we want to put a roving vehicle on the Moon!'" Romano, a second-generation Italian, was keen to ensure that, should a vehicle be commissioned for the Moon, it would have a GM badge on it!

The Army's concept at this time was a classic military vehicle, with a large pressurised cabin to ferry soldiers and equipment around. The full metal design, complete with caterpillar tracks to cope with possibly deep, loose dust areas, looked similar to the tractors of the early 20th century. It would weigh 900kg, be powered by rechargeable batteries, and have a range of between 80km and 240km.

Romano's bosses were keen to be part of this new initiative, and decided that he should recruit someone who knew more about off-road mobility. They recommended Dr Mieczyslaw 'Gregory' Bekker, a professor at the University of Michigan who worked for the Army's Vehicle Locomotion Research Laboratory. Bekker had a vast amount of knowledge about how vehicles move through different soils and he had recruited a small team of talented engineers. One of his team was a Hungarian refugee called Ferenc Pavlics.

Pavlics received his technical education from the University of Budapest in the early 1950s, but after the 1956 Hungarian revolution he and his wife fled to the US. On arrival they were interned in Camp Kilmer, New Jersey. Pavlics recalls that he'd only been in the camp for five days when Dr Bekker showed up and interviewed him. He hired Pavlics and four other men on the spot, giving each $5 and a train ticket to Detroit.

Along with the rest of the team, Pavlics began working on soil mechanics; the science of how materials react under pressure. He studied the engineering properties of surfaces, including their bearing strength and cohesiveness. The team investigated how military and agricultural vehicles maintained traction and mobility instead of sinking into the soil. They created soil bins, the vehicle engineer's equivalent of the wind tunnel, to develop a scientific methodology for relating vehicle mobility to different surfaces. In these bins they tested a variety of scale models in order to determine which ones worked best under different conditions.

Romano at GM was quick to hire both Bekker and Pavlics to continue this sort of work. "They were very excited at the prospect of building something for the Moon," remembers Romano. Devising a vehicle to traverse the lunar surface fascinated them. The challenge provided a very difficult off-road situation; not least because so little was known about the Moon's surface at the time.

LEFT Dr Mieczyslaw 'Gregory' Bekker; a professor at the University of Michigan in 1960, who also worked for the US Army's Vehicle Locomotion Research Laboratory. *(BIS)*

LEFT Ferenc Pavlics (left) at work in the early 1960s. *(NASA)*

LEFT Sam Romano (left) during the early 1960s. *(NASA)*

The answer to the question of what the lunar surface was made of lay out of reach of telescopes of the time. NASA's Ranger programme, an attempt to capture high-resolution

BELOW Best telescope views of the Moon at the time – 300m resolution. *(LPI/NASA)*

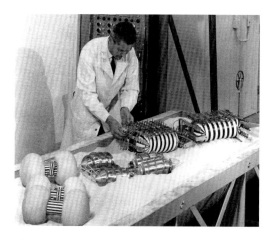

RIGHT Various small scale lunar vehicles with articulated frames and unconventional running gear. The vehicle on the left has a high width-to-diameter wheel ratio, the middle vehicle uses screw propulsion, and the right vehicle has spaced link tracks. *(Don Freidman, formerly of GM DRL)*

pictures of the surface using robotic kamikaze missions to transmit them to Earth, had so far failed. In the absence of any facts, some scientists predicted that it was covered in a very thick layer of loose dust-like material created by "the ceaseless bombardment of its surface by Solar System debris". One such scientist, a physicist called Thomas Gold, initially suggested that visiting astronauts would sink into the dust.

Based on such predictions, Romano, Pavlics and Bekker began soil bin experiments in an attempt to represent the theoretical lunar surfaces that they'd need to find a way of driving across.

"To simulate the very loose, fluffy stuff we used wheat flour," remembers Pavlics. "We had a terrible problem with mice and rats in the lab." Each morning when the team arrived for work they would find signs of rodent activity and half-eaten experiments.

Bekker's team looked at all sorts of possibilities for locomotion across such material; from caterpillar tracks to an elaborate articulated multi-wheel vehicle that would push and pull each of its segments. They even tried an Archimedean screw-type vehicle designed to burrow through the loose material in the event that the vehicle became submerged. "We tried everything," recalls Romano, "but in the end we decided that you really can't beat the wheel."

In their quest for a wheel that might be able to cope, and in an attempt to refine their soil bin studies further, they began to drop granular material (pumice or very small powdery sand) in a fairly large vacuum chamber. They wanted to see how it settled out and how cohesive (or packed) it became. They measured its bearing and shear strength and carried out so-called frictional torsion samples to find out how much traction it could provide for a wheel.

These studies confirmed that a deep dust-covered surface wasn't as soft as everyone thought. They concluded that a wheeled vehicle would not sink in very far, or have any difficulty gaining traction. So they set out to design vehicles with characteristics appropriate for riding over the surface of the dust.

In a vacuum, exposed to the direct heat of the Sun and the extreme cold in shadow, a rubber wheel would quickly perish. So Romano began to look at metal wheels, which could deal with deep dust. Under Bekker's guidance, and to make sure that the wheel could 'float' on the surface, Pavlics added metal strips to it; a feature that would inform the design of the final Lunar Rover wheels a decade later.

Project Horizon never progressed beyond a feasibility study, but its legacy would be in the people it brought together; some of whom would go on to be instrumental in persuading NASA to take a car to the Moon.

A robotic vehicle for the Moon

In 1961 General Motors' Defense Research Labs moved to Santa Barbara in California. Bekker was now head of its Mobility Research Laboratory, with Romano acting as chief of lunar and planetary programmes. In May of that year Bekker and Pavlics released a paper entitled 'Lunar Roving Vehicle Concept: A Case Study'. It was written as part of the company's contract with NASA's Jet Propulsion Laboratory for a preliminary study of an unmanned remote control roving vehicle to be sent to the Moon as part of the forthcoming Surveyor spacecraft programme.

Still uncertain of the exact size of obstacles they would have to contend with on the Moon, the General Motors team had opted for an articulated six-wheeled design. Pavlics felt that a flexible articulated frame and six wheels connected in pairs by flexible axles would have the capacity to traverse much bigger obstacles.

Building on their earlier work on metal wheels for Project Horizon, Bekker and Pavlics collaborated with the Goodyear Tire & Rubber Company to develop a new wire-frame wheel concept for the remote control rover.

Their wire weave was based on the structure of a classic 'bias tyre', in which the sidewalls and the rolling surface are interdependent (unlike a radial tyre where they are independent). The wires were looped at their intersections to maintain their spacing and still allow for what is called pantographing – a pivoting action that occurs between tyre-cords when the tyre is deformed.

Each wheel of the proposed Surveyor rover was 92cm in diameter and 38cm wide, and was driven by a 1/15th horsepower DC electric motor powered by silver-cadmium batteries. The whole vehicle was just 366cm long and 152cm wide and was configured to weigh only 30kg on Earth in order to simulate the surface load of a 182kg vehicle on the lunar surface.

Based on what the lunar surface was then presumed to be like, Pavlics selected a sand dune area in Arizona for field tests using scale models of their six-wheel rover, and the team learnt a great deal about the challenge of extra-terrestrial off-roading.

With its articulated chassis and flexible axles it gave the vehicle an elastic property that enabled it to literally climb over seemingly impossible obstacles.

The wire frame wheels also proved very well suited for a potential lunar environment because they provided excellent traction capability even for the low weight of a vehicle in lunar gravity. In addition they prevented clogging between tread pieces, since soil was free to flow in and out of the tyre.

Although the idea for a robotic rover was eventually dropped from the Surveyor programme, the pioneering work undertaken for it would pave the way for a future wheeled vehicle upon which astronauts would drive across the lunar surface.

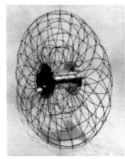
a: Wire frame wheel

b: With traction pads

c: With partial mesh cover

d: With full fabric cover

FAR LEFT Surveyor wheeled vehicle design (six-wheeled articulated) from GM for JPL. It shows the first generation of wire-frame wheels. Seen here in action climbing over obstacles in the arroyo next to JPL after delivery. *(NASA/JPL)*

LEFT Prototype wire wheel designs made by GM with Goodyear for the Surveyor Lunar Roving Vehicle. *(Dave Glemming, Goodyear/NASA)*

REINVENTING THE WHEEL

The Lunar Rover's wheel design benefited greatly from the work of two pioneers in engineering: Englishman Thomas Rickett and Polish immigrant to the US, Mieczyslaw G. Bekker. Bekker was a researcher in the Polish Army before moving to Canada in 1943 and the US in 1956. Prior to his work, most understanding of how wheels work on land came from intuition. His studies in the new discipline "terramechanics" helped the design of military and off-road vehicles and he joined General Motors in 1961 to help with their designs for possible future lunar rovers. During this time he experimented with a wide range of designs including vehicles with tracks and legs, and even a screw-propelled arrangement that would be effective if the Moon turned out to be a sea of quicksand incapable of supporting conventional wheels. In an exercise that literally reinvented the wheel, he based his designs on a 19th-century metal train wheel that had been patented in 1858 by Rickett.

ABOVE LEFT Skeet Vaughan (far right, in sunglasses) on a lunar, geology field conference in August 1965. *(Earl Roarig)*

ABOVE RIGHT A young Eugene Shoemaker with lunar surface mockup model from 1964. *(USGS/NASA)*

BELOW The Barringer Meteor Crater seen from the air, circa 1954. *(USGS)*

BELOW RIGHT Aerial view of the nuclear crater 'Danny Boy'. *(NASA)*

INFERRING THE NATURE OF THE LUNAR SURFACE

During 1964, as an environmental engineer at NASA's recently founded Marshall Space Flight Center, it became Otha 'Skeet' Vaughan's job to establish what the lunar surface environment was like. "We were interested in things like surface roughness," he recalls. "How hot it was in the Sun or [how] cold in the shade it would be. How dusty it was and how many craters and boulders there were and of what size."

Vaughan started with the lunar atlases that had been compiled from Earth using telescopes. These generally resolved features down to about 500m in size, improving to 200m in moments of what astronomers refer to as "good seeing". Their meticulous mapping of craters of different scales superimposed on yet more craters suggested that the lunar surface was probably more coherent than theories of deep dust predicted.

Combined with thermal telescopic mapping of the surface and radar signals bounced off the Moon, Vaughan tried to figure out how rough the lunar surface might be at the scale of a small, wheeled vehicle. Seeking advice, he turned to the Center for Astrogeology recently established at Flagstaff in Arizona by the United States Geological Survey (USGS).

This exotic sounding group was based right on the doorstep of one of the world's largest well-preserved impact craters, known as the Barringer or Meteor Crater. At 1,200m in diameter, it was similar in size to many of the smaller craters that had been mapped on the Moon using telescopes. The father of astrogeology and the first director of the Flagstaff facility, Eugene M. Shoemaker had mapped the crater, proving that it was made by an impact. As such, it offered insight into the debris distribution that might accompany similar craters on the Moon.

Although relatively young geologically – the impact occurred only about 50,000 years ago – Meteor Crater's debris had been badly weathered and was not sufficiently discernible to offer insight into what sort of impact crater debris a wheeled vehicle might have to contend with on the Moon. Instead, Shoemaker suggested they turn to fresher craters in the Nevada desert which had been produced by nuclear weapons. "We found some nice large craters like 'Danny Boy', which had a large amount of debris," remembers Vaughan. This and other fresh craters allowed Vaughan to map the debris concentrations and block size distributions out from the craters, giving a good idea what a Lunar Rover would have to negotiate on the Moon.

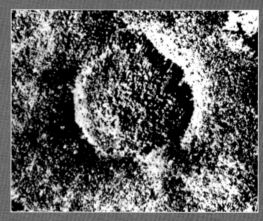

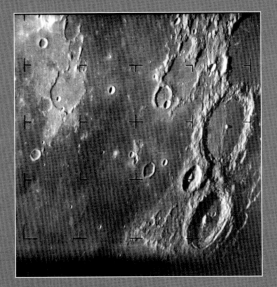
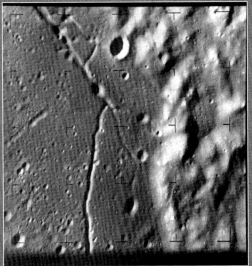

FAR LEFT Ranger 7 approach shots of Mare Cognitum. *(NASA)*

LEFT Ranger 9 views the approaching Moon at a low Sun angle to accentuate the rolling nature of the landscape. *(NASA)*

Three soil engineers on the recently formed Apollo Geology Team, Dr Nicholas Costes, Dr Dave Carrier, and Dr Jim Mitchel, devised the first informed soil mechanics model for the lunar surface. Costes applied his considerable expertise of locomotion through snowfields to their model, as such conditions were thought to be comparable to the lunar soil.

Combining all these studies, Vaughan wrote up his thesis on the nature of the lunar surface to present to his boss. He concluded that it was probably strong enough to support a lander and a wheeled vehicle. Confirmation of this new model would have to wait until much higher resolution pictures of the proposed landing sites were acquired from NASA's first successful robotic surveying missions.

Ranger, Lunar Orbiter and Surveyor

Throughout the 1960s NASA ran a series of robotic exploration programmes to better understand the lunar surface environment in preparation for the manned Apollo missions later in the decade. But the data gathered by them would also inform the design of the Lunar Rovers that would one day follow the astronauts' footsteps.

Ranger

Despite the failure of the first six missions, NASA's Ranger programme hit its stride in the summer of 1964 when Ranger 7 performed a successful kamikaze dive to a point just south of Copernicus Crater, transmitting 4,300 pictures from its six cameras. The light-coloured streaks radiating from Copernicus and from several other large craters were revealed to be chains of small craters formed by the fall of debris ejected from the primary impacts that made the large craters. To reflect the improved understanding of this part of the Moon, the International Astronomical Union called it Mare Cognitum (Known Sea). And Ranger 7's final half-frame picture, which was snapped just 600m above the surface one-fifth of a second before impact, revealed details less than ½m across; which was a resolution more than 1,000 times better than the best available by telescopes.

In February the following year, the next Ranger mission successfully swept an oblique course over the Moon, south of Oceanus Procellarum (Ocean of Storms) and Mare Nubium (Sea of Clouds) to crash in Mare Tranquillitatis (Sea of Tranquillity) near where Apollo 11 would soft land 4½ years later. It captured more than 7,000 images covering a wider area than Ranger 7, including a highland area.

The following month the final Ranger mission successfully charted the 90km diameter crater Alphonsus. Its 5,800 images mosaicked together provided strong confirmation of the multiple crater-on-crater nature of the lunar surface. And the shadows cast by the low Sun angle beautifully accentuated the rolling contours of the terrain.

From these first close-up views of the Moon further details about surface strength on a

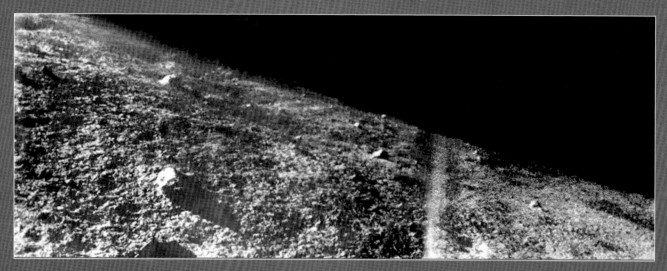

ABOVE The first ground-level views from the Moon's surface. The images taken by Luna 9 were intercepted by the radio-telescope at Jodrell Bank in the UK and published by the *Daily Express* ahead of the official Soviet release. They showed nearby rocks and a horizon just 1.4 km away. *(NASA)*

smaller, more local, scale could be inferred. The presence of large rocks on the surface favoured Vaughan's reasoning that the surface was strong enough to support both a Lunar Roving Vehicle and a Lunar Module. But *in situ* measurements were really needed to establish the maximum pressure that a spacecraft could safely exert on the surface.

Surveyor and Luna 9

Reflecting how close the race to the Moon was in the mid-1960s, both superpowers made their first soft landings within four months of each other in 1966. The Soviet Luna 9 (which was their 12th attempt at a soft landing) reached the Ocean of Storms on 3 February. Using a retrorocket to slow down, then an air bag to cushion its impact at 54kmh, the probe made the first ever survivable landing of an artificial object on another celestial body.

NASA's Surveyor 1 spacecraft also reached the Ocean of Storms on 2 June. In this case, the 294kg craft drew to a halt at a height of 3m and then switched off its engines and dropped to the surface. Images from its TV cameras included a close-up view of its footpad, revealing how little it had sunk into the dust. This photograph and the Luna 9 success further vindicated Vaughan's predictions, and

RIGHT AND BELOW Surveyor 1 surface photography showing the lander's footpad penetration into the dusty lunar surface. *(NASA)*

BELOW View of the Surveyor III footpads and the depressions which were made upon landing on the Moon. These photographs were taken during the Apollo 12 second extravehicular activity (EVA-2) on the surface of the Moon. *(JPL/NASA)*

offered direct evidence that the Moon's surface could support a heavy lander.

Subsequent soft landings continued to support this conclusion. In fact, when Surveyor 3 made the next successful lunar landing in April 1967, again in the Ocean of Storms, a failure of its descent radar prevented its engines from cutting off and it touched down, lifted off to a height of 10m, touched down again and rose to a height of 3m before a signal from Earth cut off the engines. Because the craft had come down on the inner slope of a sizeable crater, when it touched down a final time its flexible legs caused it to rebound and slip downslope before coming to rest. The fortuitous radar failure thereby confirmed a solid surface of significant strength.

Surveyor 3 was the first mission to carry a soil sampling scoop, used to dig four trenches up to 18cm deep. Samples of the excavated soil were placed in front of the TV camera to be viewed from Earth.

The results confirmed a very fine layer of dust sitting on a much more coherent surface which the scoop couldn't penetrate. Using these *in situ* measurements and observations, it was decided that the Lunar Rover's wheels should not exceed a bearing pressure on the Moon of more than one pound per square inch.

Thomas Gold revised his estimate of how far the astronauts' boots might sink into the Moon's surface to just 3cm.

Lunar Orbiter

NASA's Lunar Orbiter programme ran in parallel with their Surveyor programme. Five spacecraft orbited the Moon between August 1966 and August 1967, exposed high resolution film, scanned it and transmitted imagery covering 99 per cent of the lunar surface. The highest resolution revealed objects just 1m in size – good enough to observe the now inert Surveyor 1 sitting on the Moon.

Because the surface was imaged at low to moderate Sun angles, the Lunar Orbiter photographic mosaics were particularly useful for studying the morphology of the lunar surface. These photographs also provided the basis for the first reliable maps of small-scale surface structures and roughness information such as the slope distribution and

ABOVE LEFT Surveyor 3's trenching arm in action. *(NASA)*

ABOVE This Lunar Orbiter 2 oblique northward view towards Copernicus Crater shows slumping of the wall caused by soil liquefaction following the impact that formed the crater. *(NASA)*

BELOW A Lunar Orbiter 3 view of Surveyor 1 standing on the surface. The presence of boulders and the lack of slumping on the crater walls added weight to the theory that the lunar surface was potentially strong enough to support a lunar lander and a rover. *(NASA)*

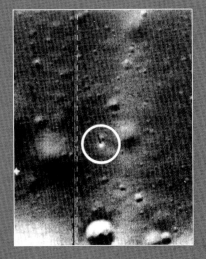
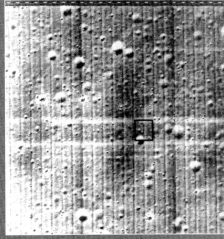

the impact crater debris patterns that any lunar roving vehicle would need to negotiate. Such information allowed the first tentative selections of possible lunar landing sites.

Skeet Vaughan used the images to come up with environmental design criteria for the rover designers to use. "We had to know something about how the [lunar] soil characteristics would affect the performance of the vehicle," he explains. He and his colleagues scrutinised the images for details such as individual boulders that had bounced and rolled down hillsides. "You could see the indentations on the boulder as it bounced on the surface of the Moon," points out Vaughan. "Even von Braun came to look at them. Such observations suggested a thin layer of something soft, and a hard layer beneath."

Field locations were sought on Earth which had similar surface roughness in order to test prototype rovers and determine how much power it took to traverse such terrain. Given the considerable cratering which these new images showed, there was also concern about how undulating the surface would be on a rover-sized scale. At certain speeds the engineers feared that the rover's suspension system could resonate with the amplitude of the terrain, causing its wheels to bounce off the ground as much as 35 per cent of the time at high speed and so make the vehicle difficult to control. It was felt that a lunar vehicle's speed should be limited to under 10kmh in order to avoid becoming airborne.

Vaughan recognised that, like the Earth, the Moon also had a wide range of different types of terrain, and he needed to quantify these in terms of the rover's ability to traverse them. So in his September 1967 report to NASA summarising the results of his exhaustive study of the collective robotic surveys, Vaughan came up with a series of models for the different types of lunar terrain he'd identified; ranging from the smooth mare (or 'sea') plains, to the rougher and geologically more interesting highland regions.

To assist the engineers in drawing up the specifications for a lunar rover Vaughan defined the ranges of slope angles, soil cohesiveness, and debris distributions for each type of terrain that a vehicle might encounter. This seminal report would inform NASA's first detailed specifications of their LRV, issued in the summer of 1969 before a single human had set foot on the Moon.

But despite the comprehensiveness of Vaughan's analysis, uncertainty about local variations in the strength of the Moon's surface remained until Apollo 11. And when Neil Armstrong stepped off the LM footpad onto the lunar surface, mission planners insisted that he be tethered to the 'Lunar Equipment Conveyor', a lanyard for ferrying kit between the LM cabin and the surface, in case he sank into the dust and needed to pull himself back out.

But Thomas Gold's early fears were proved unfounded when, a few moments later, Armstrong and later Buzz Aldrin took their first walk on the Moon's surface without sinking in more than a few inches.

RIGHT Neil Armstrong tethered to the LM during his first steps on Apollo 11. *(NASA)*

RIGHT A close-up view of Buzz Aldrin's boot and footprint on the lunar surface during the Apollo 11 moonwalk. *(NASA)*

The birth of NASA's manned lunar mobility programme

In July 1960 Dr Wernher von Braun was appointed head of NASA's newly formed Marshall Space Flight Center (MSFC) in Huntsville, Alabama. Its main objective was to undertake the development of the mighty Saturn V booster rocket to carry heavy payloads into and beyond Earth orbit.

Keen to maintain the momentum of some of the concepts proposed in his Project Horizon report, and to create a significant role for his rockets, von Braun charged a young test engineer called Otha 'Skeet' Vaughan with determining more about the nature of the Moon's surface. Using all of the available evidence, Vaughan concluded that the surface was probably strong enough to support a lunar lander and a wheeled vehicle (see pages 20–24), thereby making possible von Braun's ambitious plans for lunar exploration.

Von Braun's canny move came as Yuri Gagarin became the first man to fly into space, with President John F. Kennedy responding in his famous May 1961 speech to Congress: "I believe this nation should commit itself to achieving the goal, before this decade is out, of landing a man on the Moon and returning him safely to Earth."

Von Braun's solution to Kennedy's challenge involved two Saturn V launch vehicles for each Moon shot; one carrying the crew riding on board a 'Lunar Surface Module' and a second hauling a 'Truck' housing all their equipment, supplies, and a transport vehicle. With this heavy lift solution in mind, Marshall began to engage engineering companies to explore designs for what they called the Lunar Logistical System (LLS).

LLS studies began in the autumn of 1962 by contracting the Grumman Aircraft Engineering Corporation and Northrop Space Laboratories Inc. But they weren't the only aerospace companies interested in making their mark on the Moon. Encouraged by Marshall, many of the large aerospace firms that were already involved in building the Saturn V began to work on prototypes for a lunar transportation system.

The mighty Bendix Corporation would eventually gamble a staggering $12 million of their own money on R&D for a lunar rover in the hope of winning a lucrative government commission for this part of the national space programme.

Chrysler, General Electric, Grumman, and Boeing were also all interested in leaving tracks on the Moon. Lunar locomotion experiments across the nation gave birth to a plethora of designs employing novel wheels based on discs, cones, metal bands and even springs.

By early 1963 MSFC had positioned itself as the most knowledgeable NASA centre when it came to logistical support for America's Moon programme, and so were selected to carry out an exhaustive study into the subject.

ABOVE President John F. Kennedy and Wernher von Braun tour one of the Marshall Space Flight Center laboratories on 11 September 1962. *(NASA)*

ABOVE LEFT AND LEFT A couple of designs for lunar rovers from the early 1960s. *(NASA Goddard)*

Not surprisingly, their comprehensive 10-volume report on the Apollo Logistic Support Systems (ALSS) included a giant pressurised vehicle weighing between 2,940 and 3,840kg! They concluded that this would need to accommodate two men with their consumables, equipment and instruments on expeditions lasting up to a fortnight – the length of one period of daylight on the Moon. They called it MOLAB, standing for Mobile Laboratory.

MOLAB

In early 1964, with a view to commissioning a prototype MOLAB, Marshall began a review of the lunar vehicle design being carried out across America, and progress reports were submitted on designs from Bendix, Boeing, Chrysler, General Electric, and Grumman.

By June of that year there were just two companies left in the running; Bendix and Boeing (who had partnered with General Motors' Defense Research Labs). Two contracts were awarded with identical requests to develop a prototype vehicle. These 'Mobility Test Articles' (MTA), as NASA called them, had to carry up to four astronauts in a pressurised cabin and enable them to make lunar traverses lasting up to a fortnight.

The Boeing-GM concept was an articulated four-wheeled rover that towed a two-wheeled trailer. The whole thing was 11.5m long and weighed over 3.5 tonnes. To carry this extra load, the wires for the tyres were crimped at their crossing points in order to hold them in position, and a much higher density of wires was used to build the tyres.

Each elaborate woven-wire wheel was powered by an individual electric motor and a liquid-oxygen-hydrogen fuel cell provided the electrical power.

The proposed Bendix rover, with its distinctive hoop-spring wheels was nearly as large at 9m in length, but weighed in at a staggering 10 tonnes. Both vehicles could travel the required 90km.

The exercise had established several crucial design elements that would later pay dividends for NASA. These included Bekker and Pavlics' novel metal wire wheels incorporating an electric motor inside each; a vital concept for which the team would receive a patent in 1969.

ABOVE Early MOLAB design, including concepts from Bendix. *(NASA/MSFC)*

RIGHT Prototype wheel developed for MOLAB by GM and Goodyear. *(Dave Glemming, Goodyear/NASA)*

BELOW Joe de Fries, of the Aero-Astrodynamics Laboratory, displays the wheeled MOLAB model, and behind him is the transport configuration to carry it to the Moon using the Saturn V rocket. *(MSFC/NASA)*

Both the Boeing-GM and Bendix MOLAB designs had delivered most of what was asked of them, but both vehicles were far too big to be useful to the slimmed-down Apollo programme.

Scaling back

In November 1964 Robert C. Seamans, NASA's associate administrator, announced that all the Apollo Logistic Support Systems research would be shelved, including MOLAB. As he later joked, "Kennedy had only tasked us with landing a man on the Moon, not providing him with a car."

As part of these cuts, the Lunar Excursion Module (LEM) was renamed Lunar Module (LM) to indicate that it would not offer capabilities for powered 'excursions' away from the landing base camp, and there would be no separate shelter. Apollo crews would have to live inside the very cramped LM.

Balancing dwindling budgets against the need to generate a science capability from Apollo, Marshall pressed on with a slimmed-down research programme into what it now called the Lunar Scientific Survey Module (LSSM). The engineers began examining unmanned robotic rovers and Bendix and Boeing were awarded study contracts to evolve their giant MOLAB designs into smaller roving vehicles.

In a highly public endorsement for the new smaller lunar transportation system, von Braun

published an article in the magazine *Popular Science* in which he referred to the new vehicle as a "moon jeep". It would be an open vehicle (resembling R.A. Smith's drawing in the 1950s) with room for two astronauts wearing pressure suits and backpacks containing their Portable Life Support System (PLSS). In this article, the idea was even raised of mounting the "moon jeep" on the LM for its flight to the Moon, with the astronauts lowering it onto the surface after landing – a concept which would eventually come to fruition.

Despite NASA's official cancellation of their lunar vehicle programme, Marshall continued their own rover research and commissioned a series of test articles to determine the mobility performance of a MOLAB. To this end, these MTAs were built without pressurised cabins in order to have the same weight on Earth

LEFT Artist's concept for the Lunar Scientific Survey Module (LSSM), one of two designs for a Lunar Roving Vehicle (LRV), depicted on the lunar surface. A Bendix Corporation concept, this configuration weighed more than 8,000lb, was 21ft long, 15ft wide and had six wheels that were 5ft in diameter. *(NASA/MSFC)*

LEFT Slimmed-down Bendix concepts. *(NASA/MSFC)*

ABOVE Grumman's MTA MOLAB vehicle concept. *(NASA Goddard)*

ABOVE Grumman's slimmed-down MTA with its conical, self-cleaning wheels. *(NASA Goddard)*

as a complete MOLAB would in lunar gravity. Their principal role was to test novel wheel and chassis designs.

One such visionary concept came from Grumman, who were already building the Lunar Module. A classic car design uses something called Ackermann steering (see Chapter 2) where the chassis stays rigid and the front wheels on the inside and the outside of the turn trace circles of a different radius. Such a design can result in the wheels digging holes in

FLYING ACROSS THE MOON

In the early 1960s Bell Aerospace Systems had developed a rocket pack called the 'Bell Rocket Belt' for the US Army. It used hydrogen peroxide fuel to provide powerful controllable thrust. However, the pilots had to wear special insulating clothing because the exhaust gases could reach temperatures well over 700°C.

Ever keen to explore all the options for lunar locomotion, MSFC hired Bell to develop a version of this rocket pack that would enable astronauts to fly short distances over lunar terrain which was either too steep or too rough to traverse using a wheeled vehicle.

Bell produced two prototype designs configured for a 1-g environment. They were initially tethered to allow the pilots to become familiar with the sensitive controls. Untethered, the devices allowed a person to leap over small distances, cruise just above the ground at up to 16kmh, and hurdle obstacles 9m high. The designs were widely tested on Earth, with spectacular demonstrations of manoeuvrability over challenging terrain. The simplest design was a personal rocket belt, and this was featured in the 1965 Bond film *Thunderball*. A more elaborate prototype involved a platform that the pilot would stand on. There was even a tandem version to enable two astronauts to cruise the lunar landscape. Unfortunately Bell could never achieve flight times of greater than 20sec before the system had to be refuelled.

Despite this major shortcoming, there was serious support for such an airborne approach to lunar surface mobility, particularly from groups pushing for more geological science to be conducted on each Apollo mission.

NASA clearly saw the advantages of the Bell systems too, and even commissioned a sophisticated simulator to train their pilots. But whilst greater fuel efficiency would have been possible on the Moon, owing to the reduced gravity, the risks inherent in flying such devices were never going to be resolved. Complete concentration was required at all times, particularly when it came to landing. And then

ABOVE Bell's "rocket belt" design for lunar exploration, modelled by Astrogeologist Eugene Shoemaker. *(USGS)*

there was the danger of running out of fuel whilst in flight, resulting in a potentially fatal fall. NASA eventually abandoned this form of locomotion, seeing it as an unnecessary extra risk for the astronauts.

Bell's rocket belts did support astronauts on at least one mission though, on the set of the television series *Lost in Space*!

the ground as they turn, kicking up rocks. In an attempt to overcome this problem Grumman's MTA rover was steered by bending its chassis at a pivot point in the middle, allowing the driver to veer in the direction he wanted to go without the wheels actually turning and digging in.

Combining this articulated chassis steering system with a cunning conical fibreglass wheel that rolled any debris it collected straight back out again, Grumman created an off-road rover that was second to none in terms of its handling characteristics.

GM continued to develop their wire wheels for an MTA vehicle, creating their largest yet with a diameter of 152cm. Off-road tests at the Army's Yuma Proving Ground in Arizona did not go well however: high speed impacts with the rocky terrain broke multiple wires and a slow crawling test through a boulder field permanently dented the tyres. Nevertheless, these demoralising results would lead to important adaptations of GM's design; including a titanium inner frame and arranging tread strips in a chevron pattern around the outside of the wire tyres to aid traction.

As part of this Boeing post-MOLAB study, an exhaustive comparative evaluation of candidate lunar wheels was undertaken. Eight designs were considered, including the Bendix hoop-wheel and the GM wire-wheel, and, since they were commonly used on Earth, rigid-rim and pneumatic tyres. Features such as mechanical reliability, soft ground performance, ride comfort, environment compatibility, steerability, and weight were considered; with each category being awarded a score in the range 0-10 and then summed for an overall score. The GM wire-mesh wheel rated highest overall by a considerable margin, with the Bendix hoop-wheel coming second.

New improved versions of both designs were subjected to further intensive testing in the lab on two different types of soil chosen to resemble the surfaces seen in the pictures provided by the Ranger missions. Both rolling and steering resistance were studied and the wheels' structural limitations were measured through random obstacle impact and endurance testing on a smooth surface for up to 75,000 revolutions.

Such stringent tests pushed both types

LEFT The initial 101.6cm diameter wire mesh GM wheel (left) and Bendix hoop spring (right) wheels evaluated through the mid-1960s. *(NASA Goddard)*

of wheel beyond breaking point. The springs in the Bendix hoop-wheel started to fail after 18,000revs, with more than half being broken by 28,000revs. GM's wire-wheel faired rather better, with just under 90 per cent of its wires still intact at 27,000revs. Two further iterations of each wheel were built and tested, with both designs eventually exceeding the endurance tests and coming close to the design requirements for the eventual LRV (see Chapter 2).

In parallel with this wheel study, three significant studies of human interactions with rovers also continued to be funded through the mid-1960s.

The first also involved the Astrogeology branch of the USGS, which was seeking help in refining fieldwork procedures for human exploration of planetary surfaces. To do this, during 1964-5 GM built what they referred to as a Lunar Mission Development Vehicle (LMDV). This was a petrol powered vehicle 5.2m long that weighed 2,272kg. It rode on pneumatic tyres 137cm in diameter and had a top speed of 40kmh.

The second study on human factors, called Lunex II, was undertaken by Honeywell in

BELOW GM's Lunar Mission Development Vehicle. *(NASA Goddard)*

ABOVE Mike Vacarro (NASA MSFC) and Haydn Grubbs (Brown Engineering) suited up for a human factors exercise in the MOLAB Simulator at the Marshall Space Flight Center.
(NASA Goddard)

RIGHT Wernher von Braun driving a small Lunar Rover Mobility Test Article built by the Brown Engineering Company (BECO). It was also used for human factors studies and mobility evaluations. The development of this vehicle began around June 1965 for an in-house study of a fixed-geometry, least-risk Lunar Scientific Survey Module (LSSM) type vehicle to determine its feasibility.
(NASA Goddard)

Minneapolis during the spring of 1966. This simulated a lunar surface expedition inside a static MOLAB vehicle interior mockup that was just 10ft long and 7ft wide, and was intended to fine-tune cabin interiors, work schedules, and crew living conditions. The deputy chief of design integration at Marshall's vehicle system division Mike Vacarro, and his co-pilot Haydon Grubbs Jr., entered the chamber at the end of February but weren't told when they would be let out again. Each day, the pair followed a highly structured 14hr work schedule in which they 'drove' the static tube over a simulated course viewing through a scope on their instrument panel.

In an attempt to simulate a proper lunar expedition the pair lived off low-residue freeze-dried food, and their bodily waste was carefully collected and measured. Each day they would don pressure suits and life-support backpacks and walk for 10-15min on treadmills, investigating the 'terrain' for 'planted' rock and mineral samples supplied by the University of Minnesota. Mission trainers even simulated emergencies in which the crew had to retrieve a mannequin representing an astronaut in trouble, using the simulated airlock to bring him inside! When the simulation came to an end after 21 days, von Braun turned up to greet and thank Mike and Haydon in what he referred to as "the can opening".

"We were good friends at the beginning," recalls Vacarro, "but we weren't too friendly towards the end." Personalities aside, he did concede that they could have gone on for longer.

The third study that would help to inform the ultimate designs of the Lunar Roving Vehicle was carried out by the Brown Engineering Company (BECO), which was based in Huntsville near MSFC.

Marshall hired BECO to assess the ergonomics of pressure-suit clad astronauts interfacing with power, telemetry, navigation, and life-support equipment on a rover. A central purpose of the study, monitored for NASA by Richard Love, was to compare the merits of a large MOLAB type vehicle against those of a smaller unpressurised "lunar jeep" type of rover.

Commercially available components were used to build a small rover. The BECO rover incorporated a crude copy of GM's wire-wheels, built with 4ft diameter inner tubes wrapped in nylon ski rope. Each wheel was driven by a dedicated electric motor, powered centrally from a series of standard truck batteries. Road tests began in early 1966.

A test area on the back lot was set up to simulate a variety of wheel-surface conditions, with craters and rock debris, and it quickly became apparent that a small rover was easier for an astronaut in a pressure suit to operate for the proposed Apollo missions. Analysis of the smaller rover's tendency to bounce and even overturn at higher speeds was also studied, both on the test track and in reduced gravity conditions on board a KC-135 flying parabolic trajectories. The results showed the importance of having very soft wheels and suspension in order to keep the rover upright and under control. Further comprehensive testing was carried out during 1966 at the US Army's facilities in Arizona and Maryland.

Although no one knew it at the time, these ongoing research programmes would one day ensure that a small rover for Apollo could be designed, tested and constructed in record time, should the call from NASA ever come. Continuing studies into the concept of a Moon car kept teams of experts together at the various contractor companies across the nation. There was no guarantee that their designs would ever come to fruition, but they hoped their work was not in vain.

Persuading NASA

Through the mid-1960s the wider science community and the aerospace companies continued to push for dual Saturn V launches for each Moon shot. They argued that the Lunar Scientific Survey Modules (LSSM) were critical to successful scientific exploration by each Apollo mission.

In a compromise concept, Grumman and Bendix independently came up with interesting manned/unmanned hybrid rover designs for what they called a 'dual mode'. It would give NASA an extended mission for the same price. "After the guys have finished on the lunar surface, they hit the button, they get in the Lunar Module, they take off, Houston takes over and starts driving the vehicle from Earth," explains Leo Junen, the project engineer for the Grumman dual-mode programme in 1967. "You could take it down [into] some craters you wouldn't have the guts to go down if it was manned!" Monitored via a live colour TV camera feed, Grumman devoted considerable resources to making their system work with the 3sec round trip communications delay to the Moon.

General Motors had a more pragmatic approach to persuading NASA to carry their vehicle to the lunar surface on a single Saturn Apollo mission.

In late 1967, with their own funding, Sam Romano's small lunar vehicle concept team at GM's Defense Research Labs started its own study into the possibility of creating an open jeep-sized vehicle which could carry two astronauts up to 60 miles across the lunar surface. Crucially their work focused on the need to fit the Rover into a section of the existing Lunar Module's lower stage which they

ABOVE **A Grumman vehicle being tested at the Yuma test facility.** *(NASA Goddard)*

BELOW **Grumman's dual-mission remote control rover tests.** *(NASA Goddard)*

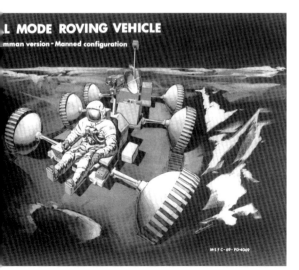

knew was empty. "It was like putting a jeep-size vehicle into the back of a station wagon!" recalls Ferenc Pavlics, who led the study.

An immense 'origami' challenge followed, but by June 1968 they were ready to go public and published a paper describing a design for a small folding Lunar Rover that would fit into Quadrant 1 of the LM's descent stage.

It was an ingenious solution to the reality of transporting a Lunar Rover on board a single Saturn V Apollo flight. But Pavlics' elegant design was overlooked by NASA. The agency was focused on winning the race to the Moon. In 1968 news that the Soviets were intending to fly a manned circumlunar mission before the end of the year prompted a last minute change of plans for NASA. Objectives for their Apollo 8 mission were rapidly altered to send it on an even more daring flight into lunar orbit. All other distractions, including plans for a Lunar Rover, were put aside.

Confident that America would still get to the Moon first, thereby opening up the need for their folding lunar rover, Romano took several of his engineers and travelled to Washington to meet with NASA. "I wanted to know exactly how much space was available in the Lunar Module, and what weight it could be," he recalls.

"Their answer was clear. The only space available for a rover would be in the one remaining empty compartment of the descent stage of the Lunar Module, to the right of the ladder [when viewing it from in front]. They said it was a space of 5ft by 5ft by 3ft," Romano recalls. To make the problem worse the shape was not square; it was triangular. "It was shaped like a piece of pie! 'What about the weight?' I asked. NASA replied, 'We can give you just under 500lb [227kg].'"

Armed with this information and knowledge of the time constraint of having to deliver a rover in time for the last Apollo flights, Romano flew back to the lab in California to tell his team. It would need to carry two men plus their equipment up to 60 miles. "We'd gone from an 8,000lb vehicle to a 500lb vehicle," he remembers thinking. "It seemed to me pretty much impossible!"

The lunar team at GM labs was down to about six people, and Pavlics remembers the mood. "We were very enthused about the possibility. But we certainly did feel the enormity of the problem because of the weight limitation. And the stowage area limitation was almost prohibitive. That's why NASA had given up on it."

They worked for four months on a design. "We had to really work some problems," says Romano. Exploiting Pavlics' 1968 ideas for a folding vehicle, they created a collapsible chassis and designed it so the wheels, when stowed, would stick out at a 45° angle in order to fit into the triangular envelope. It was the individual transmissions for each wheel Pavlics had come up with in the early 1960s that permitted such a tightly folded design. The need to fold the vehicle prompted a move towards drive-by-wire technology for controlling the front and rear steering.

But it was not enough to just pack the vehicle into this small space. "We also had to solve the problem of how to get it out," points out Pavlics. "So we came up with the solution of a step-by-step unfolding. First unfolding the centre chassis, then releasing the spring-loaded hinges to fold out the rear chassis and finally the forward chassis. Then releasing the suspensions so the wheels fold out to their final position. Then repeating the same process for the rear chassis. All this had to be accomplished with minimum assistance by the astronauts, so all the hinges were [eventually] made spring loaded."

The end result was an ingenious configuration. "[But] it was very difficult to visualise this folding concept," remembers Pavlics. "So I decided to build a [one-sixth] scale model in my spare time. It was a family affair. My wife helped to sew the seats, and I used the G.I. Joe toy of my seven-year-old son to be the astronaut!" Conveniently Hasbro, the makers of G.I. Joe, had brought out a silver space suit for the figure and Pavlics made good use of it in selling his rover concept! "My son wasn't too happy about it [at the time]. But later on he was very proud he was part of the project."

Pavlics put motors in the wheels and incorporated a remote control so that he could drive it. "It worked pretty well!" he grins, 40 years on in relating the story. To complete the pitch, they shot a little movie of the model in action, showing how it folded up and deployed.

They even shot a slow motion sequence of it traversing some rough ground to simulate its bounce in one-sixth lunar gravity.

In March 1969, just four months before Apollo 11 would make the first attempt to land on the Moon, they travelled to the Johnson Space Center in Houston, Texas to present the concept to NASA. "We didn't get a very warm reception," Pavlics remembers. "They were so focused on getting a man on the Moon, the last thing they wanted to worry about was a lunar rover!"

Undeterred, they went to the Marshall Space Flight Center in Huntsville where, after winning over the senior engineers, they trooped up to von Braun's office and placed their Rover model on the floor in the corridor outside. Using the remote controls the team stood out of view and then drove it in through von Braun's door.

Romano and Pavlics take up the story. "He was on the phone at the time. He was very surprised to see something like that coming to his desk. He ended his call quite quickly and in his thick German accent said, 'What have we here!' So we talked to him about our idea and how it was possible to do it with the existing Apollo landing system. After about 30 minutes von Braun slams his fist down on the table and shouts, 'We must do this!'"

Pavlics was aware that without a dual launch for each Apollo mission, von Braun had reluctantly given up hope of a wheeled vehicle for the Moon shots. "So he was very pleased that maybe it can be done, and this is a feasible way of doing it. He was very enthused and happy about it."

The GM folks went back to their hotel beaming. They had a big dinner and a couple of drinks and headed back to Santa Barbara the following morning expecting something great to come out of it.

However, instead of receiving a call to commission them to build their innovative folding Rover, von Braun announced the establishment of a Lunar Roving Vehicle Task Team at Marshall to request proposals from industry for a vehicle to fly with the existing Apollo landing system. "We were not surprised," says Pavlics. "That is the standard way of operating for NASA. Even so the idea was ours and we were pretty sure we were going to win!"

LEFT The one-sixth scale model that Pavlics made. *(Duncan Copp)*

A new competition

Despite Pavlics' confidence, there was going to be stiff competition. They were likely to be pitching against contractors who were already working with NASA on Apollo, notably Grumman and Boeing. Despite their vehicle expertise and novel solutions for the LRV, they lacked any manned aerospace manufacturing experience. In order to improve their chances they decided to team up with Boeing, as they'd first done on the MOLAB challenge five years before. "We decided they would build the frame – the chassis – and we would do all the things that moved," says Pavlics.

They decided to write the proposal jointly and Romano took half of his team, just three colleagues, over to Boeing. "We were met with some very surprising looks," he remembers. "Their team had over 50 people on it – talk about overkill!"

Von Braun invited Saverio 'Sonny' Morea to become the Lunar Roving Vehicle programme

LEFT Sonny Morea, NASA's MSFC LRV manager. *(NASA Goddard)*

manager. Morea had first worked with von Braun on the guidance systems for the Army's Redstone ballistic missile. And for the past six years he had managed the development and manufacture of the F-1 rocket engines for the first stage of the Saturn V and the J-2 engine for its second and third stages.

Morea still remembers taking the call. "I get a call from von Braun. He wanted me to head up the rover team. So I tell him, I don't know the first thing about cars let alone a lunar rover! But you can't say no to a man like that. I think von Braun is the best salesman the US has ever had. It's very difficult to argue with him. So I just had to say, 'OK! Yes sir!!'"

A few days later, at the beginning of May, George E. Mueller, associate administrator for the Office of Manned Space Flight, announced the Lunar Roving Vehicle as the means by which Apollo astronauts would traverse and explore the Moon. Before they'd placed a single footprint on the lunar surface or even flown the Apollo 10 'dress rehearsal' mission, the announcement of a lunar rover was a bold declaration of NASA's confidence in the future success of Apollo!

On 11 July, just days before the launch of Apollo 11, the LRV Task Team at Marshall issued its Request for Proposals for the Rover and circulated it to a number of engineering companies.

Twenty-two specific requirements were noted in the statement of work in addition to addressing the needs for deployment, mobility, controls and displays, crew station requirements, electrical power supply, and thermal control.

Without any input from direct human experience on the Moon, these initial Rover specifications were based entirely on data provided by the Surveyor landers. Insights from astronauts who'd actually been there would be factored into these engineering specifications as they returned to Earth.

NASA's July 1969 Statement of Work noting the specific requirements for the LRV (scanned from the original RFP document in the Marshall Space Flight Center archive). *(NASA Goddard)*

```
5.2  VEHICLE REQUIREMENTS

  1. Overall Requirements

     The basic requirements of the Manned Lunar Roving Vehicle to be
     procured by this statement of work are:

     1. Configuration - The LRV will be a four-wheel vehicle powered
        by storage batteries with each wheel powered by an electric
        motor. The LRV will be operated manually by one astronaut.

     2. Weight - 400 lbm maximum which includes the tie-down and
        unloading systems.

     3. Cargo Carrying Capacity - 100 lbm of science experiments
        plus two astronauts at 370 lbm each for 840 lbm total or
        alternate of one astronaut plus 470 lbm, and also to provide
        the capability of carrying 70 lbm of lunar soil and rock
        samples.

     4. Range - The LRV will be capable of performing four 30km tra-
        verses in a 78 hour period for a total of 120km.

     5. Life - The LRV will be capable of an operation life on the
        lunar surface of a minimum of 78 hours during the lunar day.

     6. Stowage - The LRV will be capable of being stowed in one bay
        of the Extended LM. The CG and the envelope of the LRV must
        be consistent with the constraints outlined in the LM inter-
        face exhibit of this statement of work (Ref. Exhibit 6).

     7. Speed - The fully loaded LRV will be capable of a sustained
        velocity of 16km/hour, on a smooth mare surface as defined in
        Exhibit 1. The LRV speed shall be continuously variable from
        0-16 km/hr.

     8. Deployment - The LRV will be capable of being deployed with
        minimum activity by one astronaut.

     9. Sterilization - Not required, but the contractor shall
        indicate his approach to reduce the level of biological
        contamination to be consistent with present LM requirements.

    10. Obstacle Negotiation - Step obstacle 30 cm high with both
        the wheels in contact at zero velocity, crevasse capability
        of 70 cm wide for both wheels at zero velocity.

    11. Slope Negotiation - The fully loaded LRV will be capable of
        climbing and descending slopes of up to 25°.

    12. Single-point failures - The LRV system and subsystem design
        will be such that no single-point failure shall abort the
        mission and no second failure shall endanger the crew.

    13. Operation - The LRV will be capable of being checked out
        and operated by one astronaut on the lunar surface with
        the controls and displays located on the vehicle.

    14. Crew Safety - The LRV design and the LRV operational pro-
        cedures shall include the required provisions to insure
        crew safety from all identified hazards. (Examples of
        hazards are solar glare from reflecting LRV surfaces,
        lunar surface roughness, vehicle instability, etc.)

    15. Reverse - The LRV will be capable of backing up with
        provisions for the driver to have visibility when
        operating in this mode.

    16. Dust - Critically affected surfaces or components shall
        be designed to minimize degradation by dust and should
        be located such that dust coverage is difficult.

    17. Clearance - The LRV will be capable of a minimum ground
        clearance of 35 cm on a flat surface.

    18. Lateral and Longitudinal Static Stability - Minimum pitch
        and roll angles of 45° with full load.

    19. Turn Radius - Approximately one vehicle length.

    20. Emergency Aids - Emergency aids will be considered to help
        free the vehicle. (e.g., hand holds)

    21. The power system shall provide a contingency 150 watts
        over and above the LRV requirements while driving.

    22. The contractor shall specify the LRV acceleration capa-
        bility in the proposal.

  2. Mobility System

     a. Introduction and Summary

        This section of the statement of work defines requirements
        pertaining to the vehicle mobility system. This system
        shall include the chassis, the flexible wheels, the drive
        or transmission, the drive motors, the suspension, and the
        steering subsystems. The contractor shall optimize each of
        these subsystems from the standpoint of weight and
        performance.

        The LRV mobility system will consist of four individually
        powered flexible wheels attached to a rigid or semirigid
        chassis frame. The chassis must accommodate one primary
        crew station, and space for all specified scientific
```

The final entries

As the drama and jubilation of Apollo 11 unfolded in front of the watching world, contractors around America scrambled to get their designs for a Lunar Roving Vehicle together. They had just 60 days to respond. On 23 July, with Apollo 11 still on its way back to Earth, a bidders' briefing was held at the Michoud Assembly Facility near New Orleans, Louisiana where the Saturn rockets were assembled. Only four companies sent representatives: Chrysler Space Division, the Bendix Corporation, Grumman Aerospace, and the Boeing-GM group.

Each company had been carrying on with their own rover studies through the late 1960s, and NASA was familiar with what they had to offer.

Apollo Lunar Vehicle 1, by Boeing-GM

Working together with Henry Kurdish, Boeing's Rover project manager, Romano and Pavlics fleshed out the design for their bid. Their neat folding concept required the front and rear quarters of the aluminium chassis to hinge over the central section, with the two seats folding flat. By employing unequal lengths of the independent suspension arms, the wheels could be forced into their stowage space when folded.

Once deployed, the front-wheel steering was engaged through a rack and pinion linkage in an Ackermann style using an electric motor that deflected them at a rate of 15°/sec. The steering motors were located below the seats, and were activated by simple on/off switches in response to a joystick control handle being tilted to the left or right.

The control system would hold a steering

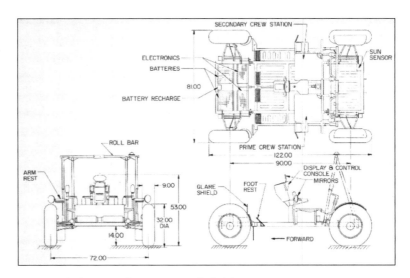

commanded turn until countermanded. A lever at the left edge of the centre chassis section could be used to crank the steering shaft manually if the motor failed.

Boeing even included a roll bar in their submission plans, accurately imagining what a lively ride lunar off-roading would prove to be.

Apollo Lunar Vehicle 2, by Bendix

The Bendix design was somewhat ahead of the Boeing-GM concept, in that they'd actually built a 1-g version of the vehicle that resembled the one they were proposing for the Moon. Sporting an evolved version of the original hoop-spring wheels they had taken forward after MOLAB, their new 416lb lightweight rover could already be driven around. This advanced prototype neatly demonstrated the capabilities of a single stick control lever to operate both the motor drive and the Ackermann front-wheel steering through two push rods. A spare portable life-support system backpack was stowed on the right, under the vehicle controller. The suspension system facilitated compact folding

TOP Boeing-GM's 1969 entry for the LRV. *(SAE)*

BOTTOM LEFT Field testing in the Cinder Lake Crater Field in September 1969 of a LRV concept built by Bendix, with future Apollo 17 astronaut Harrison Schmitt right and an unidentified person in the right seat. *(NASA Goddard)*

BOTTOM RIGHT Bendix's entry for the 1969 LRV competition. This operational mockup was used to evaluate driver boarding, PLSS exchange procedure, the deployment of simulated scientific experiments, and driving characteristics. A motorised version with a simplified crew station, two-step vehicle control, and the later steering gear was successfully driven by a subject in a hard space suit over a random course. *(NASA Goddard)*

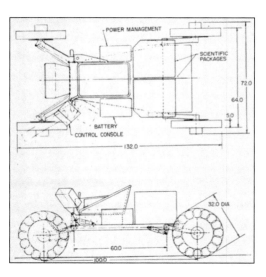

RIGHT Bendix's entry for the 1969 LRV competition. *(SAE)*

FAR RIGHT Bendix LRV folding and stowage in the LM. *(SAE)*

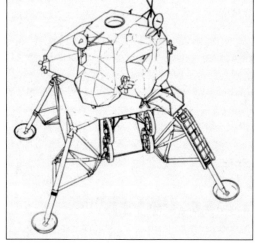

RIGHT The single track of an articulated type of steering minimises drag in rutting soils by having the back wheels follow the front wheels. *(SAE)*

BELOW Grumman's 1969 entry for LRV request. *(SAE)*

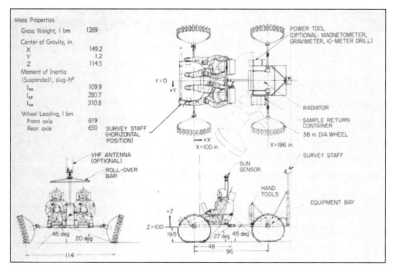

for stowage on the LM. Whilst slightly over NASA's 400lb target weight, it was a very strong contender.

Apollo Lunar Vehicle 3, by Grumman

The New York based Grumman Corporation, who were already building the Lunar Module, thought they were in with a good chance. Their novel fibreglass-epoxy conical wheels were designed for traction and floatation, and their unique articulated chassis, developed for their earlier Marshall Mobility Test Article (see top of page 28) had a lot going for it. At 379lb, it was the lightest entry – an important factor in its favour – and it boasted some novel design elements which made it superior in other ways.

The fore and aft chassis modules of riveted box girders pivoting around a flexible connecting beam offered nifty steering and also helped to keep all four wheels on the ground. Mike Vacarro, who had been involved in the MOLAB field trials a few years before, put it through its paces. "Driving over the simulated terrains my choice was the Grumman one. It was low slung. I thought it handled superior to the other designs," he recalls.

In addition to this superior handling, the Grumman rover's simpler articulated chassis steering design was also thought to be more reliable, because its fixed front axle eliminated linkages, knuckles, and other moving parts whose failure could completely disable an Ackermann steering system. As a further steering backup, if the articulation steering motor failed, this could be disengaged and the vehicle steered by differential speed control of the individual wheel motors.

Owing to their robust nature and the larger body of practical experience of them Grumman also chose a classic planetary gearing mechanism, rather than a more exotic harmonic system, for the individual wheel motors (which is still used in most cars gearboxes today).

Heat generated by all the electrical systems was radiated away through the top surface of the rear module. Heat pipes (sealed tubes containing a fluid that was readily vaporised and a capillary material) carried the heat from its sources to a temporary sink of wax just under the radiator. This melting wax alone could absorb 75 per cent of the heat from one planned excursion; sufficient capacity to ensure safe operation even if the losses through the radiator dropped to only 25 per cent of nominal.

On top of this impressive design, Grumman's knowledge of the Lunar Module, which the Rover would have to fit into, was second to none. "We certainly knew how the vehicle had to be packaged in order to hang on to the LM and be delivered down to the surface," boasts Leo Junen, a Rover project engineer at Grumman. "We came up with a very nice two-chassis deployment system, which just acted like a ladder going down to the surface."

Apollo Lunar Vehicle 4, by Chrysler

Chrysler's design sported Ackermann steering and sprung metal wheels developed from an early NASA concept. Interestingly, it was the only design to feature back-to-back seating for driver and passenger. Special transformers with bifilar windings offered a higher torque, and motor speeds of up to 2,000rpm for better initial acceleration and manoeuvring to negotiate obstacles. The driver controlled the power through a speed control lever connected mechanically to four 'trim pots', one for each wheel. By flipping a switch, forward motion of the control handle would also reverse the vehicle. A separate T-bar handle between the driver's knees controlled the front-wheel steering. Power assist for steering was accessed through thumb switches on the steering bar, which sped up the outside wheels during a turn. Computer simulations predicted that the vehicle would be able to travel at up to

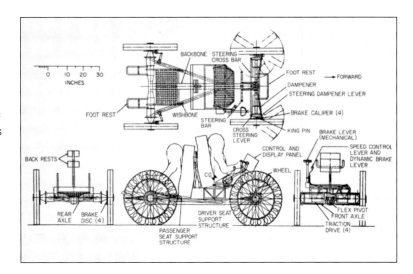

almost 30kmh on the relatively smooth mare surface, and just under 1kmh in the much rougher upland regions.

Its two silver-zinc batteries were rechargeable, and would be topped up after each sortie by twin 200W solar arrays. Heat generated by the electrical systems was to be dissipated by boiling water at just 10°C and a pressure of 0.2psi, releasing steam through vent valves. Finally, the backbone of the front chassis neatly telescoped into the tubing of the rear chassis for compact stowage on the Lunar Module.

NASA's choice

Each entry was ingenious, with a pedigree of meticulous R&D stretching back almost a decade. But this was no ordinary car contest. In the end, it boiled down to how well the vehicle could fold into the space in which it would be carried to the Moon. Boeing envisaged the vehicle deploying a bit like a "pull-down bed on wheels" with elaborate spring-loaded hinges that would minimise astronaut energy and time expenditure. And whilst the Grumman rover featured more robust performance characteristics, it required far more astronaut time and energy to assemble, making it less practical given the limited consumables and time the crews would have on the surface.

By the end of September 1969 Marshall had two contractors who they favoured: Bendix and Boeing-GM. Both were technically very close, but one was cheaper. "Both companies had come in with a bid much lower than we actually thought the rover would cost," recalls Morea.

ABOVE Chrysler's entry for the 1969 LRV competition, showing back-to-back driver and passenger ride configuration. Titanium coils and the tread of the large wheel provided suspension, and T-bar steering was achieved through mechanical linkage to the front wheels. *(SAE)*

The Bendix rover had a price tag of $22,957,000, for which they would deliver a vehicle with a target weight of 180.9kg. Perhaps without a full appreciation of the challenge, the Boeing-GM baseline price undercut Bendix at $17,280,000 with a target weight of 181.6kg.

Morea suspected that the Bendix design was a bit more costly because they probably understood more about the problem, having already built their 1-g trainer. But although not built yet, the Boeing-GM design was good too. Marshall sought permission from NASA Headquarters to get both contracts signed to legally bind each company into their pledged prices.

Then other factors were considered; such as manufacturing capability, experience of the management team, ability to meet performance goals and schedules, and even the attitudes of the individuals while making the presentations and conducting the negotiations. "Given the tight deadline, frankly I would have leaned in the direction of the fellow who was a little bit further ahead in his design," Morea admits. But in the end it was up to NASA.

Morea still remembers presenting the final evaluations to Administrator Thomas O. Paine. "He asked me up front, could either contractor do this job in the time we got? Based on the evaluations that we'd made we felt that they could, so that was the answer we gave him. So he said 'Thank you' and sent us back home. In the meantime he signed the [cheaper] Boeing-General Motors contract."

On 28 October 1969 Marshall announced that Boeing would be awarded the LRV contract, worth $19.6 million, for a promised delivery of the first vehicle to the Kennedy Space Center in April 1971.

The decision came as something of a surprise to those in the industry, and was a bitter blow to Bendix, who were relying on winning the contract to recoup some of the $12 million that they had invested in lunar transport research.

The poisoned chalice

Winning the LRV contract was perhaps not necessarily the victory that it seemed. Now the Boeing-GM team had to build what was, to all intents and purposes, a new spacecraft for the price that they had specified and to an unprecedented deadline. Previous space systems for human use had typically taken three to four years to accomplish. In fact, the Apollo space suits had taken five years!

A 17-month deadline for the LRV was considered by many people in the industry to be unrealistic. As Romano noted 40 years on, "To say we were up against it is an understatement."

Either the budget or schedule had to give. And with the Apollo launch schedule pretty much locked down, it looked like it would be the budget that would suffer.

Morea knew that a prime cause of cost and budget overruns was the introduction of design changes as manufacturing went along. From his experience with the Saturn engines, he estimated that they had spent 40-50 per cent of their time renegotiating contracts with each technical change. It was a huge time and money sink across management and engineering teams, and so he considered it imperative that they avoid making such changes to the LRV specification. So Marshall laid down very specific performance goals in a special addition to the contract.

As an extra incentive to Boeing to deliver the LRV on time and budget, Morea promised them a 15 per cent bonus if they were totally successful meeting their cost commitments and technical requirements in delivering a rover that was successfully operated on the Moon. On the other hand, he remembers, "If they came in grossly over budget they'd receive a much smaller financial reward to cover their additional expenses; around one per cent."

Actually, Morea wasn't too worried about cost overruns. NASA had earmarked $40 million for the Rover programme; the equivalent of around half a billion dollars today. What did concern him was missing the deadline for Apollo 16, at that time scheduled to be the first of the advanced missions that would use the vehicle. So he added a further, rather severe clause into Boeing's contract, stating that should they miss the deadline they would lose all bonuses, which would result in a potential financial loss for the company. Not everyone in NASA management appreciated Morea's unconventional approach, but it would pay dividends for Apollo in the long run.

ABOVE Apollo 16 commander John Young stands in front of LRV-2 in the morning sunshine of the Descartes Highlands. *(NASA)*

Equally unconventionally for an engineering project, the vendor would have to commit to the design of so-called 'flight hardware' (to go to the Moon) whilst still in the test phase of prototyping it. "We were building flight hardware assuming the prototype was the right design," observes Morea. "If you make an error you have to go back and obviously fix that problem, but it allowed us to have flight hardware just about complete when our testing was being completed, so we were able to deliver flight-qualified hardware in time."

Such a work flow was only achievable thanks to the lunar transportation research and development programme, pushed forward by von Braun since the days of Project Horizon. Without the body of knowledge this work had amassed, and the tried and tested designs that a decade of ingenuity had produced, Morea's fast-track LRV manufacturing method wouldn't have been possible.

Despite this strong knowledge base, to accomplish the task in hand, in the time available, was still daunting; even for a man who'd just presided over the manufacture of the most powerful and sophisticated rocket engines ever to fly. "It would seem like one would be considerably easier than the other," reflects Morea, "but because of the time constraints it was not so. Time, plus the sophistication of the Lunar Roving Vehicle itself, made it a very difficult task and one that kept me up many nights worrying because we just didn't have the time to go through all the testing that we needed to go through before we committed to a design."

Despite its slimmed-down appearance, the LRV was very complex, with eight primary engineering systems: mobility, electrical power, thermal control, navigation, communications, crew station, crew controls, and a flawless deployment mechanism (see Chapter 1).

Each system required built-in redundancy. And it all had to be engineered to function on rough, heavily cratered, undulating, unpredictable terrain that was blanketed in abrasive dust, exposed to the harsh vacuum of space and to temperatures ranging between +125°C and −150°C depending on whether the vehicle was in sunlight or in shadow.

The finished vehicle would be capable of travelling at 16kmh for up to 120km, of climbing a 25° slope on loose material whilst fully laden, and be stable on a 45° slope without overturning.

Every nut and bolt used in its construction had to satisfy the strict manned spaceflight requirements, which all of the other Apollo systems had spent years in development to achieve. The Lunar Rover teams at Boeing and GM would have just 17 months. At times they must have almost wished they'd not won the contract. But the engineers embarked on the project with relish. Hard work aside, designing a car to be driven on the Moon was as appealing to an adult's imagination as it was to a child's.

Work began on an Apollo Lunar Roving Vehicle in November 1969. Not even the president who'd called on his country to "take longer strides" to the Moon could have predicted where this great new American enterprise he'd initiated back in 1961 would lead.

"The visible part of the Moon extends over an area twice the size of the United States – and the far side of the Moon is just as large. As there are no superhighways on the Moon, all vehicles must have cross-country capability. Just as on Earth, the terrain on the Moon is partially smooth and flat, while other parts are rugged and mountainous."

Wernher von Braun 1964

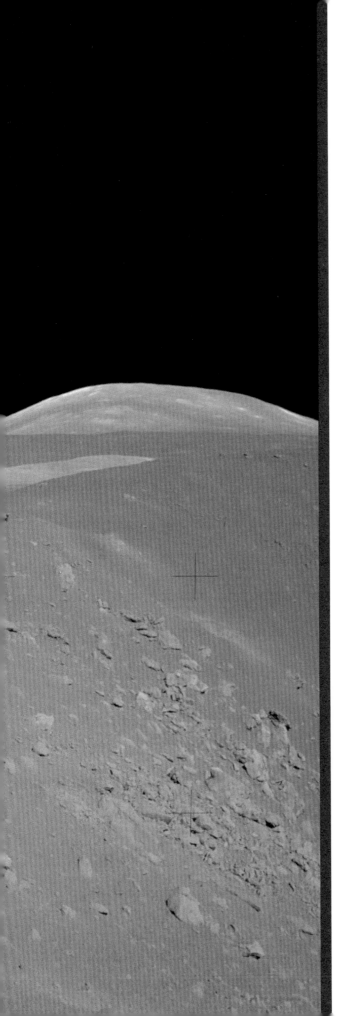

Chapter One

Structure

Seventeen months of punishing round the clock days lie ahead for the 600-strong team of engineers at Marshall, Boeing and General Motors.

The 10ft long Lunar Rover will not be going to the Moon if it can't fold neatly into an awkward triangular space just 5ft tall, 5ft long and 3ft deep. And if the astronauts on the lunar surface cannot effortlessly deploy the vehicle, then it will be worse than useless.

OPPOSITE Over 4km from their Lunar Module, Gene Cernan and Harrison Schmitt stopped at Shorty Crater. This panorama, compiled from images taken by Cernan, shows Schmitt at LRV-3 on the rim of Shorty. Nearby, just below the large boulder, are hints of the orange soil Schmitt would discover at this site. *(Gene Cernan/NASA/David Woods)*

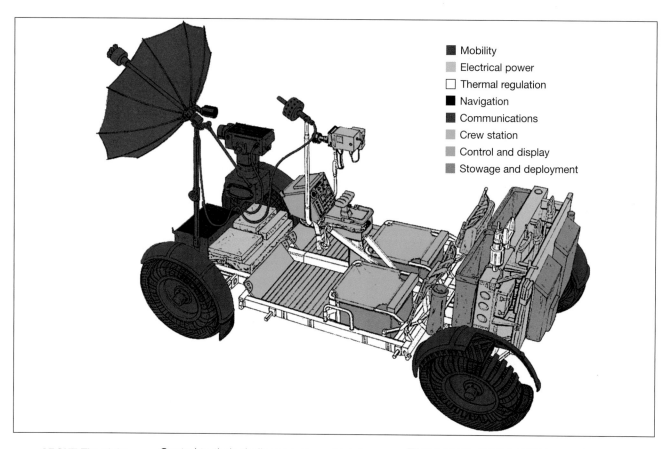

ABOVE The eight LRV systems: mobility, electrical power, thermal regulation, navigation, communication, crew station, control and display, and stowage and deployment. *(NASA)*

Central to their challenge is the vehicle's unique chassis. Not only will it have to collapse for stowage in the cramped compartment of the Lunar Module, but NASA have specified that it must also be strong enough to carry twice its own weight in men, life-support apparatus, tools, samples, and science equipment.

No wheeled vehicle has ever been built to such exacting specifications. It is a daunting prospect, and one which threatens to derail the dream of driving on the Moon once more.

Eight engineering systems

Despite its superficial resemblance to a 'dune-buggy', the Apollo Lunar Roving Vehicle was anything but simple. It had eight engineering systems, all of which were interconnected and interdependent:

- ■ MOBILITY – the largest system – consisting of the Rover's chassis, wheels, suspension, steering, motors, transmission, and brakes.
- ■ ELECTRICAL POWER – the batteries, wiring harnesses, connectors, circuit breakers, switches, and meters.
- ☐ THERMAL REGULATION – special paints and surface coatings, heat sinks, thermal straps, multi-layer insulation, and the latent heat wax tanks and space radiators to maintain optimum operating temperatures for the electronics and power systems.
- ■ NAVIGATION – distance and bearing recorders enabling constant calculations of position relative to the Lunar Module.
- ■ COMMUNICATION – the Lunar Communications Relay Unit (LCRU; stored in the Lunar Module and installed on the LRV after deployment), Ground Commanded Television Assembly (GCTA), an S-band high gain antenna to relay live TV camera pictures and a low gain antenna for voice communications and transmitting space suit and biomedical telemetry to Earth.
- ■ CREW STATION – folding seats, seat belts, folding foot rests, hand controller and arm rest, inboard and outboard handholds, toeholds, floor panels, fenders (covered in this book in Chapter 2), and stowage for tools, experiments and rock samples.
- ■ CONTROL & DISPLAY – centrally mounted display panel showing navigation (including a

RIGHT The one-sixth scale model of the proposed LRV design in 1968 to illustrate the stowage and deployment from the Lunar Module. *(NASA/MSFC)*

Sun-shadow device), speed, vehicle attitude, electrical power controls for batteries, drive and steering controls, and system temperature indicators.
■ STOWAGE & DEPLOYMENT – a series of torsion spring hinges, ropes and pulleys to release and help unfold the vehicle from the Lunar Module onto the lunar surface.

The vehicle's tight production schedule demanded that all eight systems be prototyped and manufactured concurrently, even though changes to one could potentially impose major modifications on others. Nowhere was such an uncompromising work flow going to be more challenging than with the design of the chassis. Although formally included in the Rover's mobility system, it stood apart as the single, largest component upon which everything literally hung.

At 120in long, the chassis was twice the length of the space available to store it inside the Lunar Module, so in effect it had to be folded in two. The solution proposed by Ferenc Pavlics, GM's chief Rover engineer, in his June 1968 paper (noted on page 32) was to fold the two ends of the chassis 180° so that they rested flat on the central section. He'd then tucked each traction drive and accompanying wheel under the chassis at a 45° angle, using a hinged suspension arm linkage.

This folding formed the shape of a "triangular piece-of-pie", as his colleague at GM Sam Romano was fond of saying. The flattened chassis formed one side of this triangular wedge, and the wrapped-over wheels made up the other two sides (see photograph above right). Stowed like this, the critical systems such as the drive motors, T-handle controller, crew station and display panel were all tucked inside and well protected during the flight to the Moon.

RIGHT LRV-1 folded for loading onto the LM for Apollo 15. *(NASA/MSFC)*

It was one thing for the Rover to fold into a space half its length. But it was quite another to design a method for two astronauts in pressure suits on the lunar surface to unfold and deploy the vehicle with the absolute minimum effort in a very short time.

Boeing's idea was for a series of hinges, springs and latches to drive the automatic deployment of the vehicle straight onto the surface. And as a measure of its importance, this was the first system that the engineers tested on the programme towards the end of 1969. Although they proved it could work under a variety of simulated one-sixth gravity conditions they were never able to get it to perform exactly the same way twice. But there was no time to fix it before it was demonstrated to NASA.

One hundred and twenty people showed up for the Rover's first major design review meeting at Marshall on 28 January 1970. The engineers had had just ten weeks to nail down the details of all eight systems before presenting them that day. Astronauts John Young and Charlie Duke, who were then set to be its first crew to drive the rover during Apollo 16, were also there to give their opinions.

Six of the eight Rover systems were approved, and work began on the design engineering for these systems. However the critical deployment mechanism and the proposed navigation system were both rejected for being too complicated and therefore, NASA felt, prone to failure.

This complexity worried Boeing too. Their contract to deliver the Rover to the Moon ready to drive could potentially lose them millions of dollars if any of these systems were to fail.

More seriously, should the deployment system malfunction during the rigours of launch, the partially deployed rover could prevent the Lunar Module from being extracted from its mount on the upper stage of the Saturn V. Worse still, premature deployment could damage or possibly even destroy the rocket. Everyone was worried about the vibrations of launch triggering such an event. So it was back to the drawing board, with just six months to find a solution, before NASA's next review of the design.

Weight and strength

Whilst Boeing wrestled with the deployment design issue in-house, it had other equally pressing challenges from its subcontractors, most of whom were already running over budget and behind schedule. In addition, every company involved was struggling to meet the strict weight restrictions for the Rover systems they were working on. This put extra pressure on the chassis team to devise new ways of saving weight without compromising strength.

Every ounce of additional mass carried to the Moon meant a decrease in the available hover time for the mission's commander to find a suitable spot on which to land. And, with this in mind, the maximum weight of the Rover had been set at 400 pounds. "I was told that for each pound the rover was overweight, they'd have to sacrifice a tenth of a second of Lunar Module hover time," remembers Gene Cowart, a senior Boeing LRV engineer at the time.

To deliver the desired weight, they selected an aerospace grade 2000 series aluminium alloy for the chassis build. This offered the lightness they needed and could be precipitation hardened to a strength that was comparable to steel. Between 5.8 and 6.8 per cent copper was added to the mix to produce a so-called '2219' alloy. The resulting rectangular tubing was welded together at structural joints to form the basic frame.

BELOW An early deployment mechanism test for the LRV. *(NASA/MSFC)*

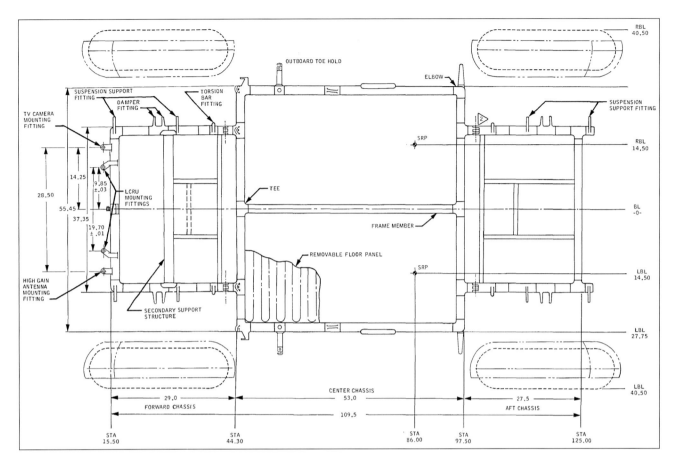

ABOVE The LRV chassis layout. Dimensions are in inches. *(NASA)*

Floor panels, also made from 2219 aluminium, served as a shear web to stiffen the frame. They were 'beaded' to reduce the tendency to flex and this permitted a significant reduction in gauge from 0.04 to 0.02in whilst maintaining the strength required to support the two astronauts and their equipment, (see illustration at bottom right of page 121).

To further reduce the weight, they milled all the main tubular chassis members down in thickness, using predictions of bending moment and shear forces as a guide for preserving vehicle strength. Such 'sculpting' was common practice on almost all Apollo hardware, but it was a battle that pitched chassis performance and strength against requisite weight.

Throughout the first half of 1970, Boeing adopted desperate methods to get the Rover's weight under control. "We tried every way in the world, even to the point of analysing it, and then testing it, and then going back and sculpting out parts of the metal to try to get the weight down," recalls Cowart. During this time he remembers turning out 1,250 technical drawings, and on one occasion 225 in one month, in a bid to control the weight.

Being over their weight limit became such a critical problem that Boeing started a contest, giving a $25 savings bond to any worker who came up with a weight-saving measure that was adopted. "You got a ball point pen for a suggestion that wasn't accepted," says Cowart, recalling a story of one very senior Boeing executive who kept sending in suggestions that had either already been tried or had been rejected as impractical. "We were fearful to go back and tell him they were no good. My immediate boss in Boeing said I'm not going to tell him his stuff's no good. So I wrote as nice a note as I could and I gave him a pen for his efforts!"

This and other endeavours to cut weight never quite reached the LRV's target of 400lb. "We might have got 400 with different materials had we had a much longer time," reflects Cowart. But eventually, with time running out, the Apollo programme managers took a hit on hover time and accepted an extra 60lb weight allowance for the Rover.

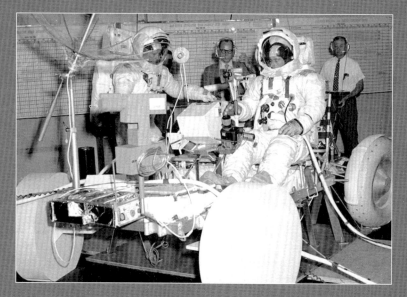

THE LRV DELIVERABLES

As if it wasn't already difficult enough to meet such a crazy production deadline, NASA had stipulated that eight test models of the rover would be built before the first of the four flight qualified units was created:

- a **full-scale mockup** to ensure that the design would comfortably "fit" the astronauts in their bulky lunar space suits and life-support backpacks
- a **mobility-design unit** to test all of the movement and control mechanisms such as motors, wheels, suspension, hand controller and drive control electronics
- a **Lunar Module unit** to determine the effects of the Lunar Roving Vehicle's weight on stresses in the Lunar Module
- **two one-sixth-weight versions** (approximating the lower gravity on the Moon) to test out the rover-deployment strategy
- a **vibration-test unit** to make sure the final product could withstand the violence of launch and the stresses of space flight
- a **qualification test unit** to be subjected to environments simulating those on the Moon. In particular this unit was exposed to lunar conditions four times longer than an actual mission, and subjected to forces much higher than those encountered during a real flight
- a **1G Rover** that was fully functional and full-weight for use on Earth (i.e. in a 1-g gravity field) to enable astronauts to train in either shirt sleeves or full EVA suits.

These were commissioned to ensure that the flight units wouldn't require costly modifications to be made later.

Built by GM's Delco Electronics Division in Santa Barbara, California, the 1G Rover had to be six times stronger than the flight versions, and hence had numerous engineering differences. In particular it couldn't be folded; the entire chassis was rigid. Nevertheless, it still had to be rated to carry almost 800lb of equipment and astronauts.

Other technical specifications for the 1G Rover can be found in the green boxes throughout the following chapters.

ABOVE Full-scale mock-up Rover to test ergonomics. *(NASA KSC)*

RIGHT 1/6th-weight Rover version for deployment training. *(NASA)*

BELOW 1G training Rover with Ferenc Pavlics, Mieczyslaw Bekker and Sam Romano at GM's Santa Barbara training facility. *(Don Freidman)*

Failure is not an option

At 02:08 UT on 13 April 1970, everyone was reminded just how exacting engineering standards had to be for human space exploration, and of what was at stake when standards slipped. A faulty thermostat in oxygen tank number 2 of Apollo 13's Service Module caused a fire that ruptured the tank, damaging other systems and crippling the spacecraft 200,000 miles from Earth. The incident threatened not only the lives of the crew, but also the entire future of the Apollo programme.

Whilst the world held its collective breath, willing the safe return of the astronauts to Earth, the team in Huntsville pressed on with their work; unflinching in their commitment to deliver a flight-ready Rover in time.

But just over two weeks later, with the Apollo 13 crew safely back on Earth, a far less public incident brought their work to a halt. On 29 April a routine static-load test on the chassis at Boeing's plant broke the frame before the target load had been reached.

Such an engineering failure was to be expected when working with such tight design margins, but with NASA's next Critical Design Review just six weeks away it spooked everyone on the programme, not least Sonny Morea, who was managing the Rover for NASA.

Beginning to doubt Boeing's ability to honour its contract, and with the entire future of the Apollo programme still hanging in the balance, Morea summoned his closest advisors to a meeting on 18 June in Huntsville to discuss the pros and cons of cancelling or continuing with development of the vehicle. "It was a very stressful period for everyone involved," he recalls. The LRV was a very high profile, prestige programme, which had been publicly promoted as being essential to Apollo's last expeditions to the Moon.

Not only did the Rover promise to maximise scientific return and hence value for money, it also offered a chance through its live TV camera to reignite public interest in manned lunar exploration. To now cancel the Rover could damage NASA in many ways and perhaps even jeopardise its future.

With such repercussions in mind, Morea's team decided to continue with the contract, with the delivery date for the first flight-qualified vehicle in April 1971; in good time for use on Apollo 16. To urge Boeing into action they reiterated the financial incentives for them if they delivered a full specification Rover on budget and schedule. Failure, Boeing was also reminded by NASA, would result in them receiving just 1 per cent of their original fee and a bruising financial loss.

After intense scrutiny, and thorough testing at Marshall's labs, NASA approved the final designs at its Critical Design Review meeting at the end of June 1970.

In an effort to meet the contracted delivery date, Boeing appointed Gene Cowart as their new chief engineer on the Rover. "They pulled me in to take over the project. And the first thing I learn is they've spent all the money and are very behind on development," he reflects. "It was clearly going to take us every man-hour available to get the Rover together."

TWO WHEELS BETTER THAN NONE

To ensure a wheeled vehicle, should Boeing miss their deadline, NASA built and tested a slimmed-down two-wheeled alternative to the Lunar Rover. This lunar motorcycle was more of a curiosity than a practical solution, but it was still tested in one-sixth gravity on parabolic flights on board a KC-135, as well as on off-road test tracks with test subjects wearing pressure suits.

The wheels, at least for the 1G version, were pneumatic tyres, and test driving proved just how hard it was to motor through a soft dusty surface on only two wheels; as anyone who's attempted to ride a bike through sand or soft snow will affirm! The Lunar Motorcycle was never used on the Moon, and the programme was apparently scrapped after the four-wheeled LRV proved a spectacular success on its first mission.

The 'Lunar Motor Cycle' developed in parallel with the Lunar Roving Vehicle during 1971 as an experimental alternative. *(NASA/MSFC)*

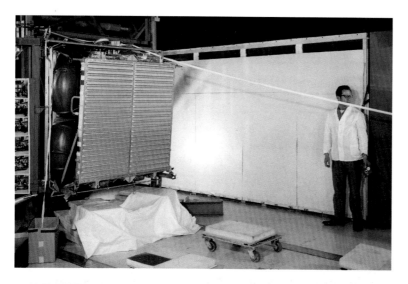

THIS PAGE Demonstration of the semi-manual deployment of the Lunar Roving Vehicle. Note the three-strut tripod stowage support near the front left wheel, one strut of which is reused as an outboard toehold (referred to in Chapter 5). *(NASA/MSFC)*

A new method of deployment

In simplifying the Rover's deployment mechanism, Boeing's intention remained for the vehicle to pop straight out when an astronaut pulled a lever, and be ready to load up and drive away.

The new concept, which was agreed with NASA at the Critical Design Review, worked like a switchblade by springing the vehicle out like a pocketknife using special torsion sprung hinges to unfold the forward chassis during deployment.

A major deployment demonstration in Huntsville was scheduled to occur in mid-1970. It would be attended by Eberhard Rees, who took over as director of the Marshall Space Flight Center when Wernher von Braun moved to NASA Headquarters following the success of Apollo 11. Gene Cowart remembers the day. "At the last minute someone pointed out that there would probably be Moon dust on everything. So we sent to town and got a great quantity of cheap talcum powder and sprinkled it all over everything." With the set dressed, the suited astronaut subject stepped forward to pull the D-ring on the side of the Lunar Module, releasing three pins to trigger its deployment. "It sort of came out," he recollects. "But it only went halfway and stopped with a loud crack. Powder went everywhere. There was a big mess. I remember Rees saying 'I don't think it will work like that,' and I replied, 'It's not going to work like that!'"

The inconsistent performance of this fully automatic deployment system led Marshall to start assisting Boeing in creating a complementary, semi-manual deployment system as a backup.

Marshall's goal was for the deployment process to take no more that 15min. German engineer Willie Prasthofer suggested the use of a ribbon, rather than a rope, and a kind of geared 'fishing reel' was developed to hold the deployment ribbon. This would lower the vehicle to the ground gradually, in a more controllable manner. Astronauts John Young and Charlie Duke contributed considerable design feedback to assist Marshall's engineers A.P. Warren and his assistant Roy Runkle with this new system.

Because a Lunar Module might end up sitting on an uneven surface, and could be tilted as much as 14.5° from the vertical, the bottom edge of its descent stage could be anywhere between 14 and 63 inches off the ground. This range of conditions was at the heart of the inconsistency problems inherent in the automatic deployment switchblade system. But Warren and Runkle's semi-automatic system proved more reliable because the procedure could be halted at any time to investigate problems that such angles or heights might present.

See Chapter 6 for the full semi-manual deployment procedure.

A change of plan

In August 1970, facing mounting challenges to its budget for Apollo, NASA confirmed that it would return to the Moon just four more times. Although Apollo 18 had been cancelled, three Lunar Rovers would still be required because Apollo 15 had been recast as an advanced J-class mission and given the first vehicle, with 16 and 17 using the next two. The fourth LRV would be used for spares (see Foreword by David Scott) and in studies of a so-called 'Dual-mode Lunar Roving Vehicle', or DLRV, mentioned on page 31). This dramatic change of plans piled further pressure on the Rover engineers, as their first vehicle would now fly to the Moon as early as

LEFT Apollo 15 mission poster from mid-1971, announcing the first rover drive on the Moon. *(NASA/Karl Dodenhoff/J.L. Pickering)*

RIGHT **LRV-1 preparations for stowing on board Apollo 15's Lunar Module at the Kennedy Space Center in the spring of 1971.** *(NASA)*

July 1971, less than a year away. Delivery by April that year was even more imperative.

The new urgency of their deadline was compounded the following month, when, under further financial pressure, NASA imposed a new upper limit on the Rover workforce of 300 people. These cuts significantly raised the workload of those who remained. "I used to say if I got to read the Sunday paper before I got a call it was a big victory," smiles Cowart.

With the bulk of the design engineering completed, including the delivery of the eight test units, and the project moving into its manufacturing phase, Cowart decided to relocate the whole Rover operation from Huntsville to the Boeing plant at Kent in Washington State. With more extensive manufacturing capabilities, and better access to the resources they needed, he felt this was the only way they would have a chance of delivering LRV-1 in time for NASA's new mission schedule.

Cowart's decision was sound, but his reorganisation did little to remedy the fact that the automatic deployment system was still failing to perform in a consistent manner. Difficult as it was to face, Boeing's switchblade design was not going to be sufficiently reliable to risk using on the Moon for Apollo 15.

Following a thorough review of the problem in November 1970 NASA ordered a redesign. After an intense few weeks of troubleshooting, it was agreed that Marshall's semi-manual backup system would become the primary deployment technique (see Chapter 6).

These last minute changes were implemented on the Qualification Test Unit by 14 December in order to undergo exhaustive acceptance tests at Boeing between January and March.

Extra chassis fittings, and the addition of fenders and science payload mounts had pushed LRV-1's weight up even further to 493.81lb (462lb for the basic LRV and 31lb for the LM support); but it was still rated to carry over 970lb of payload and operate for 78hr on the Moon's surface with a total range of 92km.

Against the odds, on 10 March 1971 Morea received a call from Boeing telling him that LRV-1 was ready for collection. Unbelievably, given the challenge, it reached the Kennedy Space Center a fortnight ahead of its 1 April deadline; almost exactly 17 months after work began.

The following chapters reveal just how close to impossible this staggering achievement really was; as engineers wrestled to create the first car capable of driving not only off-road but off-Earth!

BRITISH LEYLAND AND THE LUNAR ROVER

One non-engineering problem which arose in 1970, and was later uncovered by historians Bettyeb Burkhalter and Mitchell Sharpe in the *Journal of the British Interplanetary Society*, was an issue with the British car manufacturer Rover, based in Solihull England. In a letter to Boeing that year they raised their concern that the words 'Rover' and 'Land Rover' were copyrighted by Rover Ltd. The company pointed out "we have no foreseeable intention ourselves of producing a vehicle for use in any space programme, and therefore we are not really in competition with you, nevertheless we feel that there is some risk of confusion…" Further on in the letter, as a helpful suggestion for resolving the situation Rover volunteered that they "…cannot see that we could possibly object to such words as Lunarover."

The Grover

The 'Grover' was the name the astronauts used for 'Geology Rover', which was a rugged 1-g trainer ingeniously constructed in the summer of 1970 by Rutledge 'Putty' Mills of the Astrogeology branch of the USGS. He had

LEFT A 1971 Land Rover. *(Les Roberts)*

been asked to build it from scratch in 90 days for field training that was scheduled to start for Scott and Irwin in July of that year.

To match Boeing's LRV blueprints and pictures, Mills recalls that he sourced parts from old cars and aircraft; including a landing gear motor from a B-26 Marauder and a torsion bar from a 1960s Morris Minor that he came across in a local scrapyard. The whole build cost a mere $1,900! Mildly embarrassed, the USGS told NASA the cost was more like $20,000; but even at that price it was a bargain!

BELOW The so called 'Grover' hurriedly built by the Astrogeology branch of the USGS in the summer of 1971 for rover field training of the Apollo 15 crew. *(NASA/USGS)*

"Let me tell you, this is quite a Rover ride. It's quite a machine, I tell you! I think it would do a lot more than we'd let it."

Gene Cernan, commander of Apollo 17 while driving on the Moon.

Chapter Two

Mobility

It would be the job of the Rover's mobility systems to convert chemical energy in the batteries into kinetic energy, and to do so smoothly, efficiently and reliably in the hostile environment of the Moon's surface. The vehicle's moving parts would have to tolerate extremes of temperature, vacuum and the notoriously abrasive and tenacious lunar dust.

OPPOSITE Apollo 17 astronaut Harrison Schmitt rakes for pebbles at Station 1 in this panorama constructed from images taken by his commander Gene Cernan in the Taurus-Littrow valley. It is the first day out for LRV-3.
(Gene Cernan/NASA/David Woods)

ABOVE Lunar Roving Vehicle-1 on its Earth support frame. Note the white fenders, made from a different resin to the red fenders taken to the Moon. (NASA)

BELOW Gene Cernan's photographs here show Harrison Schmitt working at LRV-3. Compiled into a down-Sun panorama, they highlight the undulating, rocky lunar surface which the Rover must negotiate. (Gene Cernan/NASA/David Woods)

To add to their problems, the engineers would also have to endow these systems with a high degree of redundancy. And, as with everything in the Apollo programme, they would have to juggle these exacting requirements with a constant pressure to make the vehicle's components as light as possible in order to meet the strict weight budgets of their ride to the Moon. On one level these systems had to be simple but given the outlandish environment in which they were to operate they would be exotic, clever and beautifully engineered creations.

Four sealed electric motors incorporated into the hubs of each wheel would eventually do the job, taking power from two silver-zinc batteries situated at the front of the Rover. The suspension system, tyres and steering would have to cope with a world where the gravity is only one-sixth as strong as Earth's, and on which not a single inch of road had been prepared. Yet they had also to be designed to allow the wheels to fold for transport to the Moon.

At a time when the most sophisticated electric vehicles on Earth delivered milk to villages and towns in the United Kingdom and when most Earth automobiles were not known for breakdown-free motoring, this new breed of extra-terrestrial electric car would need the efficiency of a thoroughbred racing car and the reliability of a Rolex watch. Its mobility systems would allow it and its load to traverse slopes of 25°, reach a top speed of 17kmh and take two passengers nearly 8km from their lander and back. The crew's lives depended on it, and that was the fact which kept the engineers at General Motors and Boeing awake at night.

Where no wheels had rolled before

By the time the new LRV project got going in early 1970, engineers were already well versed on what to expect from the lunar regolith; the name given to the thick layer of well-compacted dust and rubble which blankets the Moon's surface. The crews of Apollo 11 and 12 had described the plains where they walked and hundreds of high quality photographs had been taken. Small experiments had been carried out on the mechanical properties of the soil and decent quantities of the stuff were now available for study on Earth.

All this combined with data from five unmanned Surveyor probes as well as the Soviet Luna 9 spacecraft which had all soft landed prior to Apollo, meant that knowledge

of the surface was sound. It had shown that although the top few centimetres were very loose and powdery, aeons of settlement had compacted the subsurface to make it very firm. The Apollo 11 crew had struggled to push their flagpole into the ground, and when they later took off from the Moon, Buzz Aldrin was dismayed to see their rocket's exhaust plume blow the flag over.

If subsequent landing sites were similar, there would be little problem with driving a Rover across the lunar landscape as long as the wheels could cope with occasional small rocks and the ubiquitous dust. The issue was whether the more interesting sites that geologists wished to visit would be similarly navigable. But unlike the somewhat eccentric locomotion designs of the earlier decades, informed by guesswork and speculation on the nature of the Moon's surface, the LRV design team were sufficiently well informed to be confident about designing a bespoke system for the job of lunar locomotion.

Wheels

The rubber-clad wheel, so familiar to drivers on Earth, has serious shortcomings as a candidate for use on the Moon, especially over the intended 78hr period of operations. The main drawbacks are the extremes of temperature that exist in the lunar environment. The unfiltered Sun heats one side of an object to temperatures over 100°C. Yet the shadowed side gets very cold, as the only heat is that reflected off the ground. If a rubber tyre wall is exposed to the black of space for an extended period, perhaps while the astronauts sleep, its temperature can plummet to well below −100°C. In lunar temperature simulations, early tyre designs were failing, threatening the exceedingly tight 17-month deadline of the project.

It would take a Polish immigrant to the US called Mieczyslaw G. Bekker to solve the problem. Bekker had become something of an expert on how wheels interact with soils, and following his guidance the engineers for GM's 1964 MOLAB design had ditched the rubber tyre and opted for an all-metal wheel that would give a similar performance across the lunar terrain

ABOVE Rubber wheels on the Moon. Though they could not withstand long-term exposure to lunar conditions, this Apollo 14 photograph shows how the rubber wheels of a hand cart used on this mission smoothed the lunar dust into ruts that reflected the Sun. *(Alan Shepard/NASA)*

(see Introduction). Further, exhaustive testing and improvements had refined this wheel through the rest of the 1960s, with Bekker's design eventually making it into Boeing's LRV.

Early in 1970, two evolved designs for his metal tyre were evaluated. One was a flexible metal wheel with sheet metal tyres, the other was a wire frame wheel with internal hoops and a solid aluminium rim. They were tested in simulations that used crushed basalt to replicate as closely as possible the conditions on the Moon.

After careful studies the second design with the internal hoops was chosen. Testing of its characteristics and further refinements continued right up to delivery of LRV-1. In March 1971 – just weeks before Boeing delivered the first flight-qualified Rover to the Kennedy Space Center – one of the wheel discs buckled during a routine structural test. Sam Romano's team at GM had less than a month to fix it. "It was a very stressful moment," he recalls. "NASA was very upset and Boeing was worried. We were all very worried."

Romano quickly discovered that, in their constant battle to bring weight down they'd machined the wheel discs too thin. Urgently requiring four new thicker wheels to replace those already on LRV-1, GM went straight to the vendor in Los Angeles and ordered thicker, heavier discs. "We needed all the four discs spun in less than a week," Romano remembers. "He revised his tools, got new material and by

RIGHT This face-on view of a Rover wheel shows its essential elements; aluminium hub, titanium inner wheel called a bump-stop, and a woven-wire outer tyre. *(NASA)*

FAR RIGHT The wire mesh surface of the Rover's wheels being woven on a special jig. *(NASA)*

BELOW Schematic of wheel with cross section showing hub and motor/transmission placement. The wheel/hub interface is also detailed. Dimensions are in inches. *(NASA)*

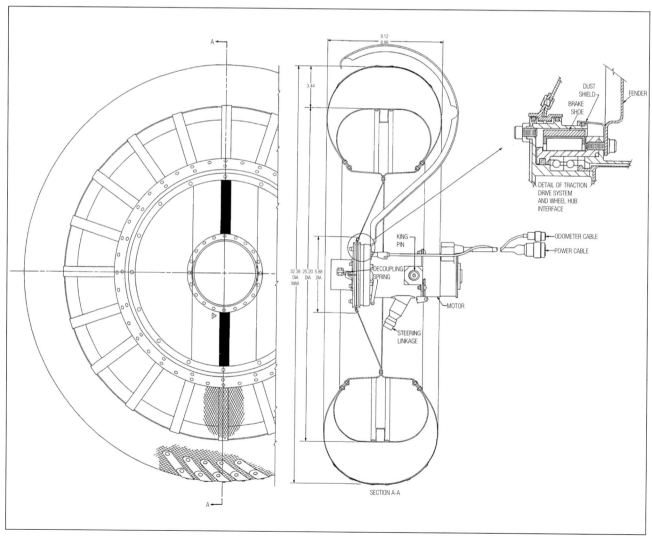

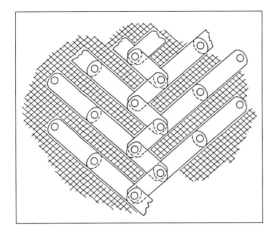

ABOVE **Detail of titanium tread strips riveted to the traction surface of the tyre.** *(NASA)*

ABOVE **Three months before Apollo 15, astronauts David Scott and Jim Irwin examine LRV-1 with space-suit gloves. The light through the left rear wheel shows the internal bump-stop clearly.** *(NASA)*

God he spun the discs for us. We built entirely new wheels and got them on the flight article just in time to get the darn things assembled and into the Lunar Module."

Throughout the stressful rush-manufacturing and delivery of LRV-1, exhaustive further testing of the wheels in simulated lunar conditions were undertaken. One of the more elaborate simulations had a full-sized wheel propel itself around a circular trough filled with powdered rock while the whole set-up was enclosed in a vacuum chamber. Once initial lessons had been learned from operating in 1-g conditions, engineers took the entire arrangement on board an aircraft. In May 1971, just weeks before the first LRV would fly to the Moon, 65 flights were conducted with the aircraft flying one-sixth-g parabolas to try and learn how much dust the wheel would throw up in the lunar environment.

Constructing these new GM-Boeing wire mesh wheels had been a labour of love for the teams of weavers involved. The outer layer of each 81cm tyre was hand woven in a special jig using 800 individual lengths of zinc-coated piano wire, every one of which had been painstakingly x-rayed for flaws. Once attached to the hub, this construction could easily deform around small rocks while still giving a ride that was similar to a soft rubber tyre. To improve the wheel's grip, titanium chevron strips were riveted to the wire mesh; these covered half of the tyre's ground contacting surface.

This outer tyre housed an inner titanium tyre, which was visible through the mesh. Known as a bump-stop, this inner wheel consisted of a ring held in position by a series of flexible bands which absorbed the impact force of collisions with larger rocks. The entire affair of wire, ring and bands was attached to an aluminium hub and mounted on the output side of the wheel's dedicated motor. Despite its all-metal construction and large size, the mass of one wheel was only 5.4kg. On the Moon, even with the slightly thicker wheel discs, each would weigh the terrestrial equivalent of just 0.9kg – less than a bag of sugar!

> *The 1G Rover tyres for primary use were pneumatic automobile tyres, but special wire mesh wheels were also available.*

The low gravity on the Moon can be deceptive because although the mass of a fully loaded Rover would be around 720kg – nearly ¾ tonne, the Moon's gravity would make this feel like only 120kg and therefore the load on each tyre at rest was only equivalent to 40kg (about 390N). However, being on the Moon does not mean it has lost any mass and on the move, it still carried the kinetic energy of a ¾-tonne mass. At its top speed of 16kmh, a collision with a rock would increase the wheel load by a factor of ten times to about 4,000N, and on Earth that would be equivalent to nearly half a tonne of weight. Concerned that permanent damage to a tyre might prevent the Rover from functioning on the Moon, thereby severely reducing their fee for the contract, the Boeing team decided to rate the tyres up to a tonne! On top of this they also guaranteed them for 120km use – over twice the expected distance.

RIGHT A set of wheels for the Moon and each lightweight assembly was only 5.4kg in mass. *(NASA/MSFC)*

BELOW Gene Cernan stands next to LRV-3 at the start of Apollo 17's third day of exploration at Taurus-Littrow. The Rover's fenders show the beading added as the engineers' tribute to the Ford Model-A. *(Harrison Schmitt/NASA)*

Despite such over-engineering, NASA was still concerned about the strength of the tyres. Boeing's chief engineer for the LRV project, Gene Cowart recalls that during one meeting the agency was worried that the wires on the wheels might break. "There was concern that broken wires might easily rupture a suit and kill an astronaut! So they set a test wheel up on a big treadmill and ran it for hours to check the wires would stand up." They did! But there were other wheel related problems just around the corner.

In the spring of 1970, dust profile tests were starting to show the need for a fender to cover the wheel. The tyre's porous mesh allowed dust to enter it when in contact with the ground and then get thrown out as the wheel continued to rotate. Early tests in vacuum conditions showed just how much dust the Rover's wheels would throw up, potentially covering the astronauts and the vehicle in dark powder. Further one-sixth-g tests on parabolic flights confirmed this problem. These dust sprays would have a knock-on effect on the Rover and astronauts because the darkened surfaces would absorb extra heat from the Sun. Since careful thought had been put into ways of managing temperatures through surface finishes and radiators (see Chapter 3), the engineers didn't want all this hard work to be undone by something as simple as the accumulation of dust, so fenders and flaps were added around the upper half of each wheel to overcome this problem. As an extra precaution against dust, the astronauts carried brushes in order to clean dust from each other and from the Rover – particularly the tops of the radiators.

In a story told to Apollo historian Eric Jones, engineer Bill Kimsey described how he and colleague Waine Borne, who both worked on the fenders, had renovated Model-A Fords as a hobby and they admired the fine beading on the Ford's fender design. Just for fun they decided to add a beading to the Rover's fenders and

LEFT LRV-2 in its folded configuration. The tendency of the wheels to bulge out when folded together is visible on the left. Before installation in the Lunar Module, retention wires will pull these bulges inwards away from the LM's insulation. *(NASA)*

they made it the same width as on the old Ford. When their boss asked them why it was there they claimed it was to stiffen the fenders; he just nodded and it was approved.

Squeezing the Rover into its temporary home on the LM was never going to be easy and the wheels were pushed against each other. To accommodate the fender, one section was mounted on rails and stowed in a telescoped fashion outside the main part of the fender. Once the Rover had been deployed on the surface, an astronaut would pull that section of fender into place; a design which would come back to haunt them in operation on the Moon (see Chapters 3 & 7).

In addition, the tyre bulge caused by squeezing the wheels against each other threatened to rub against the LM's insulation so engineers used retention wires to pull the bulge inwards towards the chassis. These wires ran from the motor casing, to the hub, the inner wheel, and finally the outer wire surface. When the Rover was deployed, these wires were released and the wheels could adopt their natural shape.

Motors

Early work by GM on lunar rover designs had shown that one motor per wheel was a good solution from the point of view of redundancy and weight, and this logic was subsequently adopted for the LRV. The maximum power output of the motors that were ultimately chosen was about ¼ horsepower each from a 36V battery supply (see Chapter 3).

Much analysis had been done on the choice of electric motor technology. A favourite of the engineers was a brushless motor that used permanent magnets whose inner workings would have been open to the lunar vacuum. However, amid concerns about the abrasive nature of lunar dust and the reliability of the control electronics, Sonny Morea, NASA's project manager, selected a more proven sealed DC brush-type motor.

The concept behind the DC brush-type motor is that electrical power is fed to a rotating armature wound with coils of wire – the rotor. This rotor sits within a magnetic field created by static windings – the stator. When current

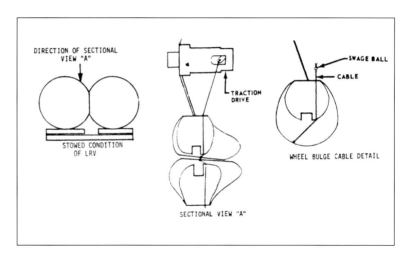

ABOVE Schematic of the bulge retention wires. *(NASA)*

BELOW Cross-section of the LRV traction drive unit. The electric motor is to the right and the harmonic drive transmission unit is to the left. Dimensions are in inches. *(NASA)*

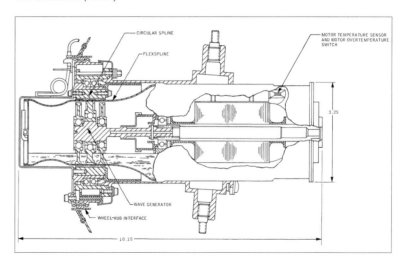

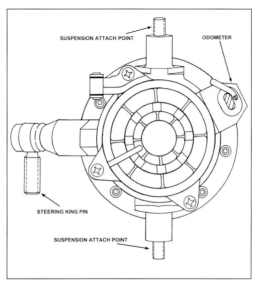

LEFT End-on view of the traction drive unit with suspension and steering attachment points. *(NASA)*

is first switched on, the rotor's coils become a temporary magnet and it turns to align its north pole with the stator's south pole. Before it can complete the task, the current is switched to run in the opposite direction and the rotor therefore keeps turning as it tries to align the magnetic polarities.

To switch the current at exactly the right time on such a motor, there is a simple rotary switch that consists of a split ring called the commutator which sits at one end of the rotor, and a set of contacts, known as brushes for historic reasons, that are pushed onto the commutator by springs. The brushes deliver the power and the commutator distributes it to the rotating coils. This arrangement goes back to the 19th century and is well developed and understood.

For the LRV motor, rather than expose its inner workings to the vacuum and dust of the Moon, it was sealed within a case with a nitrogen atmosphere at 7.5psi, about half of Earth's sea-level pressure. The nitrogen helped to distribute heat evenly around the motor. A small additional amount of water vapour was also introduced to the gas to increase the life of the brushes as they ran against the commutator.

Included with the motor were a brake drum for controlled stopping, and an odometer that generated nine pulses per wheel rotation. These pulses were used in the vehicle's speedometer and navigation system. The unit had a temperature sensor that drove a meter on the vehicle's instrument panel, and a thermal switch to protect it from excessively high temperatures. Also sealed within the case was a transmission unit which was a step-down gear arrangement with a fixed 80:1 ratio.

Motor control

To control the power and therefore the speed of a DC electric motor, it has to be supplied with a DC voltage that can be adjusted: low voltage gives low power and high voltage gives high power. An old-fashioned way to control power was to use a rheostat, a kind of variable resistor. Resistive wire would be wound around a long ceramic core and an electrical contact could slide up and down the windings. If power was fed through the length of the wire to the contact, then the resistance would depend on how far along the windings it was positioned. This is a very poor method because a lot of power is wasted by generating heat in the windings. A supply is most efficient when it is passing all of its power, or none.

In the LRV, the motors were driven by a box called the Drive Control Electronics which took this all-or-nothing approach to delivering power. The technique is called pulse width modulation (PWM). Think of it as a high-speed switch which could be cycled on and off 2,000 times per second (i.e. 2 kilohertz) as it controlled the flow of power from the batteries to the motors. By switching the power so fast, the motors could not react to the individual pulses of power, and instead responded as if being supplied by the average voltage.

If the 'on' time was equal to the 'off' time (a 50 per cent duty cycle) then the motors operated as if they were getting half the full battery voltage. If the power was on more than it was off, then more power was fed to the motors. Conversely, if the 'off' state dominated, then less power was delivered. The advantage of this was that the electronic system that feeds battery power to the wheels could operate in either of two states, fully on or fully off, both of which were inherently

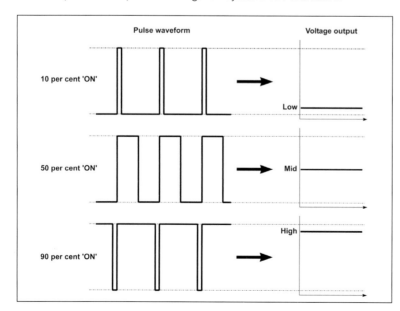

BELOW **Diagram to show the principle of pulse width modulation.** *(David Woods)*

> *The 1G Rover traction drives were not hermetically sealed.*
>
> *Direction changes on the 1G Rover could be initiated at speeds less than 1kmh without coming to a complete stop.*

more efficient than trying to generate an intermediate voltage.

When the astronaut moved the T-handle in a backwards and forwards direction, he operated a device called a potentiometer, more commonly known as a 'pot'. A very familiar example of the potentiometer is the old-fashioned volume control on a radio or TV, and it is similar to the rheostat. A resistor is formed into a circular track and this has a voltage applied across it. A movable contact, or wiper, rotates within the track and where it touches, it picks up a voltage that is a portion of the full voltage, depending on how far along the track it is touching. Unlike the rheostat, the wiper voltage does not deliver power. It is merely generating a control signal for a subsequent circuit. The wiper is mechanically connected to a shaft so that the amount of rotation of the shaft dictates what value of voltage is picked up from the track.

In the Rover, the potentiometer output signalled the Drive Control Electronics to adjust the power delivered to the motors. These electronics generated pulses at a frequency of 2 kilohertz. The width of these pulses was altered by the signal from the potentiometer. These were low-power pulses, not yet ready to send to the motors but they would act as control pulses for the next stage. To protect the motors, further circuitry inhibited them from progressing further if certain conditions were met. For example, no pulses were to be passed if the brake was on, or if the current to the motors was excessive, or if the T-handle was moving between the forward and reverse regions. If the pulses were passed, they controlled a transistor switch that allowed battery power to be fed to the motors when it was on. The Rover had to be brought to a complete halt before engaging the reverse switch and using the reverse movement of the T-handle.

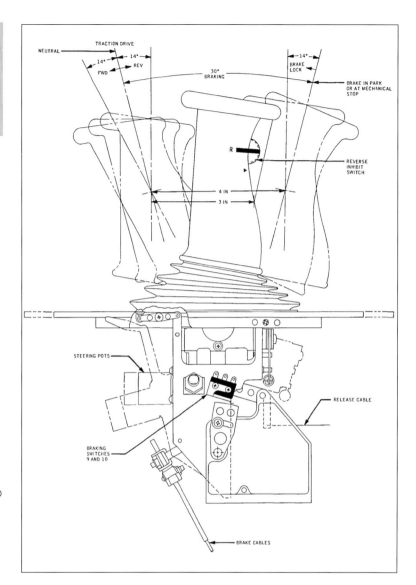

ABOVE Diagram to show movement ranges of the T-handle for motor control and braking. *(NASA)*

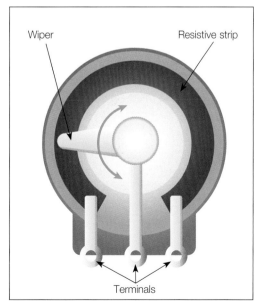

LEFT Schematic to show the operation of the potentiometer. *(Dominic Stickland)*

Transmission

On a smooth surface, running flat out, the motors could go up to about 6,000 revolutions per minute. If you directly couple that to a wheel that has an effective diameter of 70cm, it would produce a blistering top speed of over 250kmh which is plainly ludicrous. More importantly, the motor would be unable to provide the necessary torque – a measure of its ability to turn the wheel against anything that tries to prevent it turning. This includes mere friction with the soil when starting up or negotiating obstacles at slow speed. The solution adopted by most car and motorbike designs is to employ a gearing mechanism of some sort, both to reduce the speed of the motor and to increase the torque available to move the vehicle. The Rover had its own unusual gear arrangement.

Most people are familiar with gears that use a collection of toothed wheels on an arrangement of axles to step rotational speed up or down. Multiple steps must be used for large ratios which require more gear wheels, and the axles become offset which makes any casing larger, all of which adds weight. Such conventional gears are also prone to a degree of play whereby if the applied torque is relaxed or reversed, the teeth have to engage on their opposite sides resulting in a degree of slack.

In 1955 Walter Musser, an Italian engineer working in the US, came up with a clever gearbox design called a harmonic gear that was light, very compact and had a high step-down ratio. Even better, its axles were in line, it exhibited very little play, and it required only three major components. Boeing's subcontractor, United Shoe Machinery, licensed the design from Musser for use on the LRV. It was sealed in the same casing as the motor to keep it from the dust and vacuum of the lunar surface and, by placing it in a separate compartment, it could benefit from conventional lubrication with Krytox 143AC fluorinated oil that could tolerate the hostile conditions encountered in the gear.

In operation, the motor's shaft drove a device called a wave generator. This was simply two sets of roller bearings on the end of the shaft whose axes were slightly offset from the shaft's axis, each opposite the other. They turned inside a flexible cylinder called a flexspline that was fixed to the motor's casing so that it could not turn. Instead, the rollers forced the flexspline into an elliptical shape, thereby creating two so-called waves that moved around at the same speed as the motor's shaft. The peaks of the waves were merely the tips of the flexspline's ellipse. The term 'spline' refers to a cylinder or tube that has regular ridges running lengthways along its outside or inside surface. For the flexspline, its ridges were essentially gear teeth that were machined around its outer surface.

The flexspline caused major headaches for

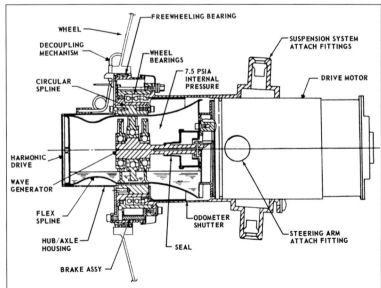

BELOW Cutaway diagram of the harmonic drive to indicate components. *(NASA/Mike Jetzer-heroicrelics.org)*

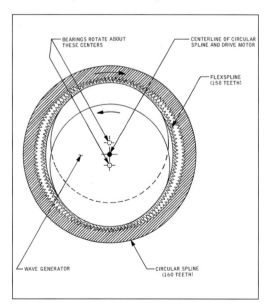

BELOW RIGHT Functional diagram (not to scale) of the harmonic drive. *(NASA)*

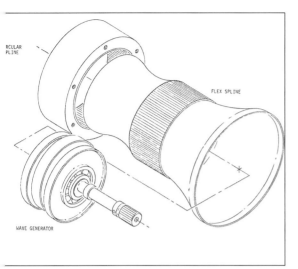

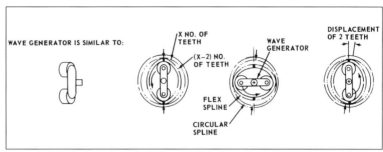

LEFT Drawing of the three major components of the harmonic drive: the circular spline, the flexspline and the wave generator. *(NASA/Mike Jetzer-heroicrelics.org)*

ABOVE Diagram to show the mode of operation of the harmonic gear. *(NASA/Mike Jetzer-heroicrelics.org)*

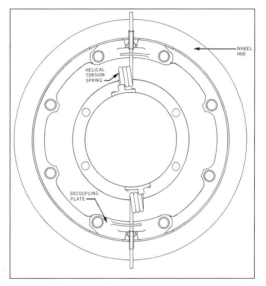

LEFT Diagram of where the wheel attaches to the transmission unit showing the two decoupling mechanisms. *(NASA)*

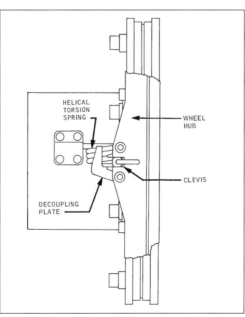

LEFT End view diagram of a decoupling mechanism. *(NASA)*

engineers during development because steel versions would crack and fail after several thousand revolutions, so the Rover team worked with metallurgists at the factories to find a metal that would withstand the harsh treatment this component would receive. It needed to be thin, lightweight and therefore flexible, yet it had to be strong to transmit power while being able to cope with the high temperatures it generated during operation. They eventually decided to fabricate it from Inconel, an alloy based around nickel and chromium noted for its hardness and durability, but which is notoriously difficult to machine, especially when 158 precision gear teeth had to be formed around the outside.

The 158 teeth of the flexspline meshed with 160 teeth on the inside of a stainless steel circular spline, but only at the tips of the flexspline's wave. The upshot of this arrangement was that each rotation of the motor's shaft and wave generator caused the circular spline to rotate by only two teeth, or one-eightieth of a rotation. Since the flexspline was fixed to the motor's casing, the circular spline was forced to rotate in the opposite direction to the motor shaft and 80 times slower.

> *The traction drive for the 1G Rover has a 3-stage planetary gearbox in lieu of the harmonic drive.*

If a motor were to seize, the wheel could be decoupled from the motor/transmission unit to allow it to rotate freely on a separate bearing.

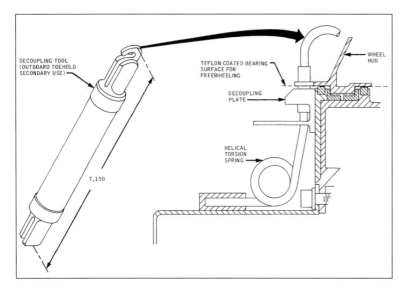

ABOVE Diagram to show the use of an outboard toehold as a wheel decoupling tool. Dimensions are in inches. *(NASA)*

This was the last stage in the transmission chain. For this, engineers included two simple mechanisms that connected the output side of the motor/transmission unit to the wheel's hub. Each consisted of a spring, one end of which was engaged in the rotating end of the motor. The other end would normally be engaged in the hub. To decouple the wheel, an astronaut could use one of the Rover's toeholds as a tool because a hook on one end could disengage or re-engage the spring as necessary.

Decoupling a wheel was not without consequences that an astronaut had to be aware of. The wheel's brake would no longer be effective and the odometer pulses would no longer be sent to the navigation system. And because the third-fastest wheel was automatically selected for navigational use, decoupling two wheels disabled the entire navigation system. In an extreme situation, Boeing figured that even if three drive units were lost (an extremely unlikely event) the remaining one could still get them back to the Lunar Module, and for navigation they could simply retrace their tracks in the regolith.

> *The 1G Rover had simulated wheel decoupling mechanisms to duplicate the LRV-to-crew interface. However, operation of this simulated mechanism did not yield actual decoupling. Furthermore, wheel decoupling on the 1G Rover did not disable the brake on the affected wheels.*

Steering

As with the motors, redundancy was built into the Rover's steering by having two separate systems, one for the front wheels and another for the rear. This really was an early form of 'drive-by-wire' as there was no direct mechanical link between the T-handle and the wheels. The connection between the two was made electronically. Each system consisted of a servomotor, gears, and linkage to the wheels. When both front and rear steering were operating, they made the Rover into a highly manoeuvrable vehicle but each could provide satisfactory steering control on its own.

The LRV's steering is of the Ackermann type which takes its name not from the inventor of the system, Georg Lankensperger who

RIGHT This view of David Scott and Jim Irwin during a session with the training rover shows one of the toeholds that could be used as a decoupling tool. *(NASA)*

FAR RIGHT LRV-3 is shown parked at Station 6 on its final drive. Its wheels are turned, displaying its double Ackermann steering geometry. *(Gene Cernan/NASA)*

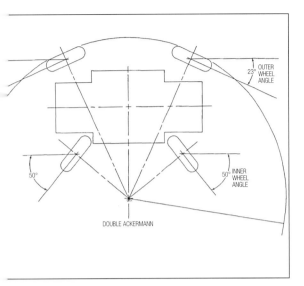

devised it in 1817, but from his agent, Rudolph Ackermann, who patented it. Ackermann steering is used on almost all modern vehicles and its essential property is that the wheel on the inside of a turn steers to a greater angle than the outside wheel in order to reflect the smaller radius of its turning circle. This way, during a turn, neither wheel is being forced to slide sideways. On the Rover, since both front and rear wheels could be steered at the same time, it benefited from double Ackermann steering which gave it a wall-to-wall turning circle of only 6.2m, only twice the Rover's length and more than one metre better than a London taxi.

The astronaut steered the Rover by deflecting the T-handle to the left or right against its spring-loaded centre position (see Chapter 5). This deflection was measured in a manner similar

ABOVE LEFT Diagram of the Rover's double Ackermann steering geometry including angles of maximum turn for inside and outside wheels. *(NASA)*

ABOVE Soon after deployment, Gene Cernan drives the unloaded LRV-3 around to the front of the Lunar Module. This image shows both front and rear wheels being used for steering. *(Harrison Schmitt/NASA)*

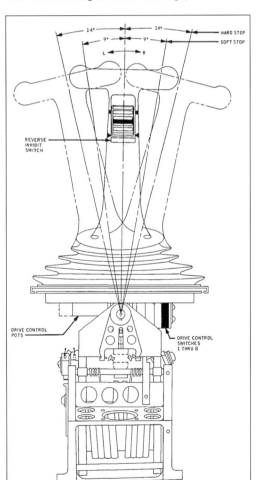

LEFT Diagram to show movement range of the T-handle for steering. *(NASA)*

FAR LEFT This unintentional shot of LRV-2's T-handle was taken by Charlie Duke when he was starting a fresh roll of film. John Young is visible beyond the brown wrist rest. *(NASA)*

65

MOBILITY

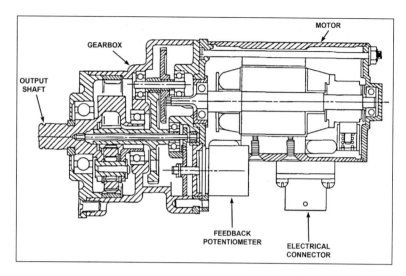

ABOVE Cutaway diagram of the steering motor and its reduction gearbox. *(NASA)*

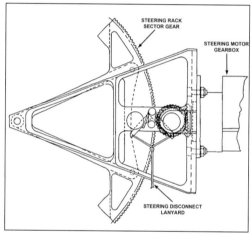

RIGHT Diagram of the Rover's steering rack. *(NASA)*

BELOW Diagram of the components of the steering linkage for the whole Rover. *(NASA)*

to speed control by using a potentiometer. The resulting voltage was proportional to the handle's movement, and it became the command signal for a servo system.

Servo systems use the concept of feedback to achieve precise control, for if an astronaut was to avoid rocks and thread between craters then his steering had to respond precisely to his commands. We all use feedback in daily life, usually without realising it. For example, when driving a car along a road, our eyes and brain sense at every moment whether we are correctly positioned in the lane or if the setting of the steering wheel is making us wander to the left or right. Essentially we have produced an error signal that represents the extent to which we are moving away from our desired road position. We feed this error signal back to our steering command by almost unconsciously turning the steering wheel in the opposite direction to cancel out the error.

The way the LRV's steering used feedback was to have a second potentiometer measure how much the wheels had actually turned. This was fed back and compared with the command signal. Any difference between the two represented an error between what had been commanded and what was actually being steered. The steering motor continuously used

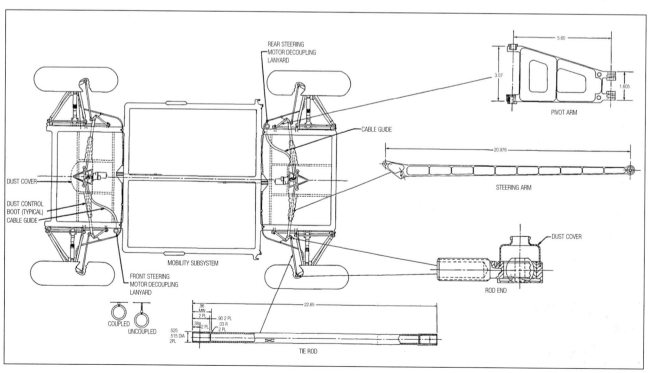

this error signal to drive the difference to zero and make the wheel direction constantly play catch-up with what was being commanded.

The steering mechanisms were located within the front and rear chassis frames and a panel on the underside protected them from damage by protruding rocks. A mechanism consisted of a one-tenth horsepower steering servomotor that ran at up to 500rpm, and a gearbox that reduced its output to move a large sector gear from side to side. This sector gear was so-called because it was only about 60° of a full circle; the arc of the 'sector' didn't need to be any greater. The sector gear operated the steering linkage which ran from the centre of the panel to both wheels.

The linkage to each wheel from the sector gear consisted of a steering arm that stretched to the edge of the chassis and linked to a pivot arm and a tie rod that took the steering movement to a king pin, the main pivot in a steering system. At the point where the steering arm and tie rod met, engineers arranged a joint that would allow the tie rod to hinge in order to enable it to accommodate the up and down motion of the suspension and allow the wheel to fold downwards for stowage in the LM. Because the king pin was directly fixed to the casing of each wheel's motor, the steering system turned the entire motor and wheel arrangement. At full steering lock, this mechanism could move the inside wheel 50° from its central position while the outside wheel went to 22°. The Rover could be steered lock-to-lock in only 5.5sec, giving a lively ride that could cause a little sliding at the back when at speed.

> *The 1G Rover steering utilised a continuously operating steering motor. Hand controller movement energised the appropriate counter-rotating magnetic particle clutches, thereby engaging the load and effecting steering. A magnetic brake was actuated when the clutches were not engaged.*

If one of the steering mechanisms failed, there was provision for an astronaut to disable either system by pulling on a lanyard that closed a scissor mechanism. This achieved two things:

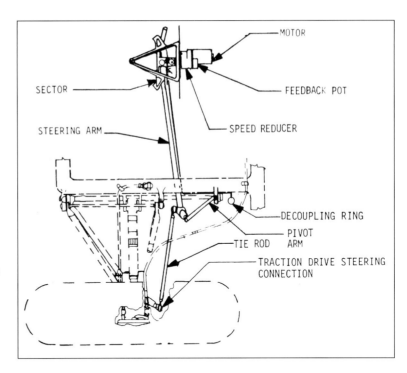

ABOVE Detailed layout of the Rover's steering system. *(NASA)*

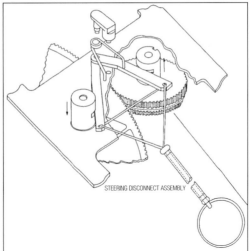

LEFT Diagram of the scissor mechanism for decoupling the steering rack from the motor. *(NASA)*

BELOW Before and after diagrams of the mechanism for disconnecting the steering. *(NASA)*

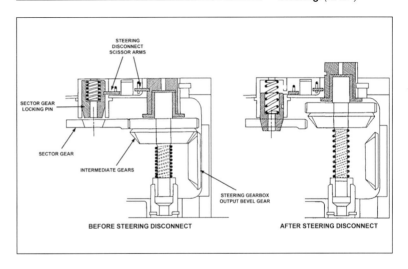

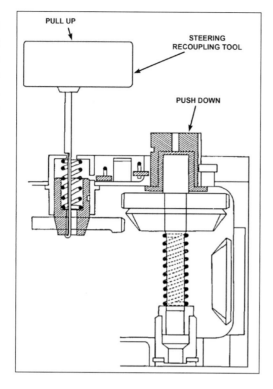

RIGHT Diagram to show how a recoupling tool can be used to reconnect the rear steering system. *(NASA)*

BELOW This panorama, created from photographs taken by Charlie Duke, shows John Young with LRV-2 on the lower slopes of Stone Mountain. The roughness of the surface is evident and was a challenge for the designers of the Rover's suspension. *(Charlie Duke/NASA/ David Woods)*

it released a spring-loaded locking pin which then engaged into the centre of the sector gear in order to lock it and keep the steering in a central position; it also allowed two gears to disengage under spring action that otherwise linked the steering motor's gearbox to the sector gear, thereby removing drive from the motor to the steering mechanism.

Only the rear steering system could be reconnected by an astronaut on the Moon because there were too many items like batteries and electronics mounted on the front chassis to allow access to the mechanism within. To reconnect, he used a tool with a hook on the end to retract the locking pin from the sector gear and re-engage it with the scissor mechanism. This tool was stowed on the rear chassis. To reconnect the steering gears, he merely pushed down on the gear axle which also latched on one of the scissor arms.

The 1G Rover had simulated steering decoupling mechanisms to duplicate the LRV-to-crew interface, but operation of this simulated mechanism did not effect actual decoupling.

Suspension

What became very important early on in the Rover's development was to fully understand the sort of terrain this vehicle was going to have to cross. Radar studies of the lunar surface provided some clues, though they were misleading and widely misinterpreted as indicating great depths of loose dust. No one was really sure that it would be possible to drive across the heavily cratered surface until super high resolution pictures taken by the Lunar Orbiter spacecraft were scrutinised (see Introduction). These revealed the surface of the Moon to be so ancient that the number of craters present at a particular scale was in equilibrium, as old structures were obliterated by later impacts. The result was a very undulating surface. Geologists also noted that fresh craters were usually surrounded by fields of boulders. Although the presence of boulders showed that the surface was strong enough to support large rocks (and hence a lander), if

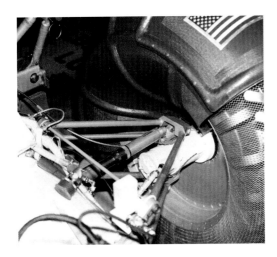

ABOVE **The suspension of the left front wheel of LRV-1. Note the upper and lower wishbone arms attached to either side of the motor casing, and the damper arm between the arms.** *(NASA)*

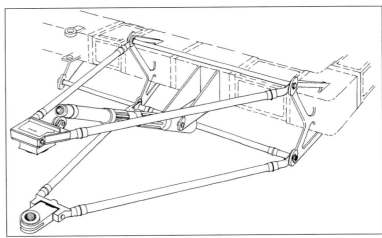

ABOVE Isometric drawing of the suspension system for one wheel of the Rover. *(NASA)*

BELOW End-on diagram of the suspension system including the range of movement during operation and folding. Dimensions are in inches. *(NASA)*

there were too many boulders they would make driving difficult.

Nevertheless, as plans went ahead for Apollo 15 and the first use of the Rover, NASA had no good images of the area that it was to visit, known as Hadley. No one was really sure of its "trafficability", NASA's term for whether the crew would be able to negotiate the landscape in their new car.

The role of the suspension system was to support the LRV's loaded chassis at a set height above the lunar surface whilst on the move. At the same time, it had to give the astronauts a reasonably smooth ride on a surface that had no prior preparation. There are no roads or paved areas on the Moon – just crater upon crater and a regolith littered with rocks ranging in size from pebbles up to boulders the size of a house. It was inevitable that the Rover would not be able to completely avoid the smaller rocks and so would have to drive over them. Though the deformation of the wire wheels would handle some of the bumps, the suspension system would provide most of the smoothing of the ride.

What made the design of the suspension most interesting was that it could never be fully tested on Earth, since it was designed to operate at one-sixth g. To simulate lunar gravity, a test rig was built into one of NASA's KC-135 'vomit comet' aircraft in order to assess its suspension and wheel performance. "It's not called the vomit comet for nothing," recalls Mike Vacarro, at that time a manned systems integration engineer. "We took von Braun up there one day. All he could take was ten parabolas before he was sick and we had to bring him down. You have to have a strong stomach. You keep those barf bags handy!" It proved difficult to properly test a car inside the cramped conditions on board this plane, so most driver training was undertaken with the 1G Rover on Earth using a specially strengthened rover vehicle with identical controls.

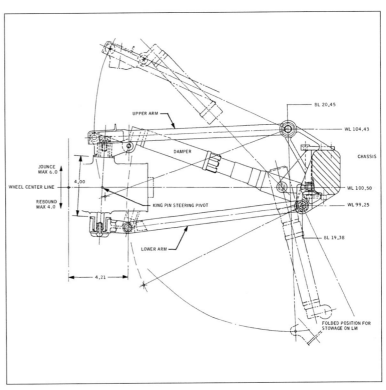

MASSIVE WEIGHT CONFUSION

How heavy is the Rover? The vehicle's specs say 460 pounds or 209 kilograms. That all seems pretty fair until we take the vehicle to where it belongs, the Moon. Now how heavy is the Rover? Since lunar gravity is almost exactly one sixth as strong as Earth, it is tempting to say the answer is 34.8kg, and sure enough, the Rover feels a lot lighter on the Moon. Yet, strictly speaking, it is still 209kg. How so?

It all depends on what we mean by 'heavy' and this has caused a great deal of confusion since the birth of spaceflight. In fact there are two concepts behind our notion of how heavy something is: weight and mass.

If we hold an object in our hand, we feel it push down with a force. This is our common understanding of the weight of something. Gravity acts on the object and makes it exert a force when we hold it in place. In other words, weight is a force; a push caused by gravity. Scientists and engineers measure force in 'newtons', appropriately named after the father of gravity himself, Sir Isaac Newton.

Another way to think of the heaviness of an object is how much matter it contains. We call this its mass. A very massive object contains more stuff than a less massive one and the standard unit for mass is the 'kilogram'. We define the kilogram in terms of a standard volume of water. The standard volume used is a cube 10 centimetres in width, height and depth. This also happens to be the definition of the litre. The water within, by definition, is 1kg of mass. The important point is that a 1kg mass is always that, 1kg, no matter where it is; on Earth, on the Moon or even floating 'weightless' in space somewhere. A kilogram is a kilogram is a kilogram.

The confusion arises because on Earth, when people think they are measuring mass, they are usually measuring weight, which is really a force. They might feel the force that gravity makes an object exert, perhaps by holding it in their hands, or they might use various types of scales. Either way, they rely on gravity and so they actually measure force, not mass. But confusingly, their scales are always marked in units of mass, like the kilogram, instead of force, like the newton. As a result, the force that an object exerts thanks to gravity gets confused with its intrinsic amount of matter, its mass. In daily life in Earth's gravity field this doesn't matter much.

The distinction between weight and mass becomes important when we face situations we are not used to, like the lower gravity of the Moon. On the lunar surface, although the mass of the Rover has not changed, its weight has reduced markedly and it *feels* lighter. In fact, if the need arose, the astronauts would lift up the Rover and reposition it.

But imagine if the Rover were to hit a large rock head-on during its travels over the lunar surface. How much damage would it cause? We have a calculation that we can do which will tell us the energy of this collision, stated in units called *joules*, named after James Joule, another British physicist. We take the speed at impact, which for the rover was typically 2.5m/sec. We square this number to get 6.25, and then we multiply it by half the *mass* of the Rover (209 × 0.5 = 104.5). The figure we get for the energy of the collision is 653 joules. And it would be 653 joules whether we carried out the collision on Earth, on the Moon or even in deep space. And so the engineers needed to make the Rover's components strong enough for its mass of 209kg regardless of its lighter apparent weight on the Moon.

Suspension systems are notoriously heavy parts of a car, but the suspension of the Rover had to be light, despite the punishment it would take on the Moon. A typical family car can carry only a third to a half of its own weight, even though its mass is typically 1½ tonnes. The LRV suspension had to carry two men, each wearing heavy suits and life-support equipment, along with an array of science equipment, tools and rock samples – altogether over 500kg. Yet the Rover itself had a mass of 209kg. Of course, the Moon's low gravity made the payload weigh the equivalent of only 87kg on Earth. Nevertheless, the suspension had to deal with the momentum of almost ¾ tonne of mass in motion over bumps and depressions in the regolith. When fully loaded, and taking into account the expected distortion of the tyres, the suspension system would also need to give the Rover about 35cm clearance with the ground.

The team under Sam Romano, GM's programme manager for the LRV, came up with an arrangement of lightweight wishbone arms, torsion bars, and fluid dampers to create a suspension mechanism for this unique job. A pair of wishbone arms, one upper, one lower, connected the chassis with the wheel/motor assembly. Each wheel had an independent suspension mechanism joined to the forward or rear chassis at pre-formed attachment points.

The connection points were all articulated, and the geometry, which was almost a parallelogram, not only maintained the wheel's approximate vertical alignment as it moved up and down in use, it also permitted the wheel to fold downwards and inwards by 45° for stowage on the Lunar Module.

The suspension of the 1G Rover was not designed to allow folding for LM stowage.

The travel of the two suspension arms was limited by a telescopic damper connected between the chassis and the floating end of the upper arm. Filled with 47cc of silicone oil, this facilitated 15cm of snap (the engineering term for the movement caused by a sharp bump) and 10cm of rebound. Heat generated in the damper was transmitted through the oil to be dissipated in the damper's structure.

Rather than use heavy coil springs, the suspension loads were carried by two torsion bars, one each for the upper and lower wishbone arms. In a torsion bar, the required amount of springiness is provided by applying the load in a manner that twists the bar. The bars ran between the support points of the wishbone arms with one end fixed to the chassis by splines (ridges that run lengthways along a bar) and the other end fixed to the suspension arm. Most of the load (85 per cent, in fact) was carried by the lower torsion bar while the upper bar was primarily responsible for powering the unfolding of the wheels during the deployment process.

Because the mechanism would be stored under compression in the salty environment at the Cape for three months prior to launch, there was concern that stress corrosion

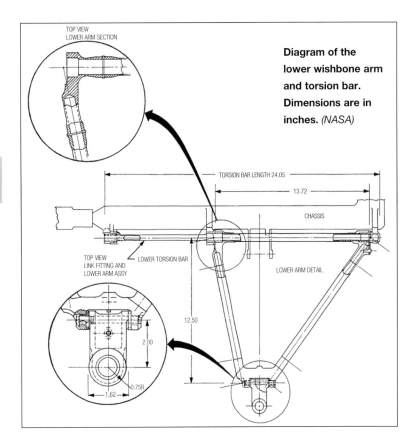

Diagram of the lower wishbone arm and torsion bar. Dimensions are in inches. *(NASA)*

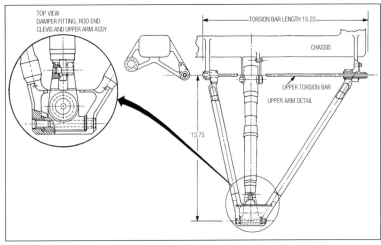

ABOVE Diagram of the upper wishbone arm and torsion bar. Dimensions are in inches. *(NASA)*

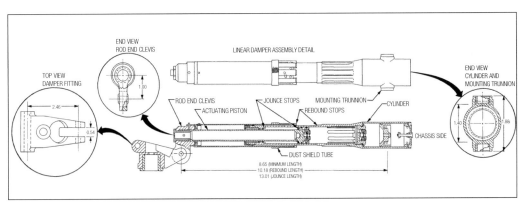

LEFT Diagram and cutaway of the suspension damper. Dimensions are in inches. *(NASA)*

RIGHT Suspension system on a replica Rover at Kennedy Space Center. *(Karl Dodenhoff)*

Because the suspension system of the 1G Rover was not designed to fold, there was no need for the upper torsion bars and it contained only the lower torsion bar on each wheel.

The brakes of the 1G Rover were hydraulically actuated disc brakes. They were actuated by the hand controller in the same manner as the LRV mechanical brakes.

might crack the torsion bars, remembers Jim Sisson, Marshall's chief Rover engineer. "And if they cracked you could not deploy it." In order to reduce the risk of stress corrosion, the materials people came up with a surface coating which also dulled their shiny exterior. "We did it overnight the day before delivery to the Cape," recalls Sisson. "And when we came to do the final review, before shipping, everyone was a bit agitated because the torsion bars were still shiny and we couldn't figure out what had happened." It turned out that the workman who had made them was so proud of their shiny finish that when he found them covered in cruddy stuff he machined it off. "They went through all that trouble to protect them and he just decided that was not the way it should have looked, so he took it off! There was a panic at the time, but looking back it was pretty funny," laughs Sisson. GM hurriedly recoated them and still managed to ship LRV-1 on schedule. "The workman just did what he thought was right."

Romano's biggest problem in manufacturing this novel suspension system was in trying to get the welds right on such a lightweight construction with little margin for over-design. They all had to be equally strong – something that proved surprisingly difficult to accomplish with the exotic materials they were using. In the end, they opted for a tungsten/inert gas welding machine at another company. The whole design 'hinged' on them getting these welds right, notes Romano. "There were plenty of long weekends. We sweated those days."

Brakes

Like any car, the Rover needed some way to slow down or come to a stop when necessary using more than the mere friction of rolling along – even though the motors provided some braking through the harmonic gear. Given

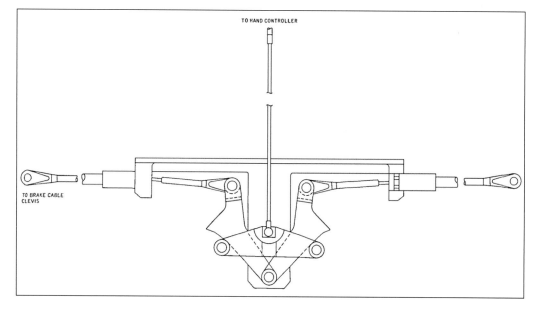

RIGHT Diagram of brake cable layout including equalisation device to deliver equal braking force to either side of the Rover. *(NASA)*

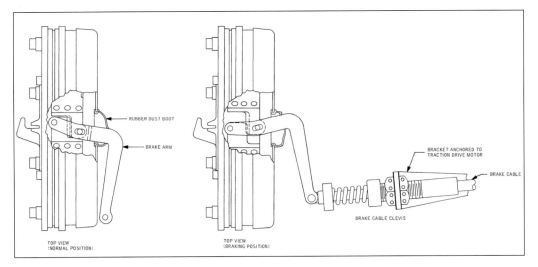

LEFT **Diagram of the operation of the Rover's drum brake. Movement of the brake lever pushes the brake band against the inside of the drum.** *(NASA)*

that the vehicle was designed to go up 25° slopes, it would have to be capable of driving downslope in a controlled fashion. Moreover, there would be occasions when the astronauts wished to sample boulders and craters on a hillside and this meant there had to be some kind of parking brake. Additionally, although the weight of the Rover and payload on the Moon was only equivalent to about 120kg, it still had a mass of 730kg, nearly ¾ tonne, and its momentum was a considerable thing to halt once the vehicle was up to speed.

The brakes were applied when the T-handle was pulled back towards the driver on the lower of its two pivots, known as the brake pivot (see Chapter 6). This operated a pair of brake cables, one for the front wheels and one for the back. Each cable then went to an equalising device that transmitted its movement to both wheels while ensuring that equal braking force was applied to each side.

The brake mechanism was a conventional drum-type arrangement very similar to a type used on many older or cheaper cars on Earth. It was mounted on the output end of the motor/transmission assembly and contained brake shoes. At the drum, the brake cable operated a lever that pushed the shoes apart to force them against the inside surface of the brake drum and apply braking force. The brake lever entered the brake drum past a flexible cover that served to keep lunar dust out.

Since the brake drum was part of the motor/transmission assembly, if an astronaut had to decouple a wheel from a stalled motor, this would also render the brake for that wheel ineffective.

This mobility system would ensure a smooth ride across the Moon's unpredictable, undulating, debris-strewn surface for a total (for all three vehicles) of almost 100 punishing kilometres. But such reliable locomotion would only be possible with an unimpeachable electrical power supply, the subject of the next chapter.

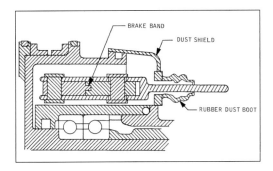

LEFT **Cross-section diagram of the brake assembly.** *(NASA)*

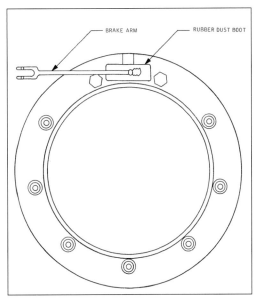

LEFT **Rear view diagram of the brake drum and lever.** *(NASA)*

"The batteries were a worry to a lot of people. We had done an extensive amount of testing on the batteries and always were confident. There's always that thing in the back of your head though, that if they don't work..."
Jim Sisson, chief engineer for the Lunar Rover, MSFC, 1969-71

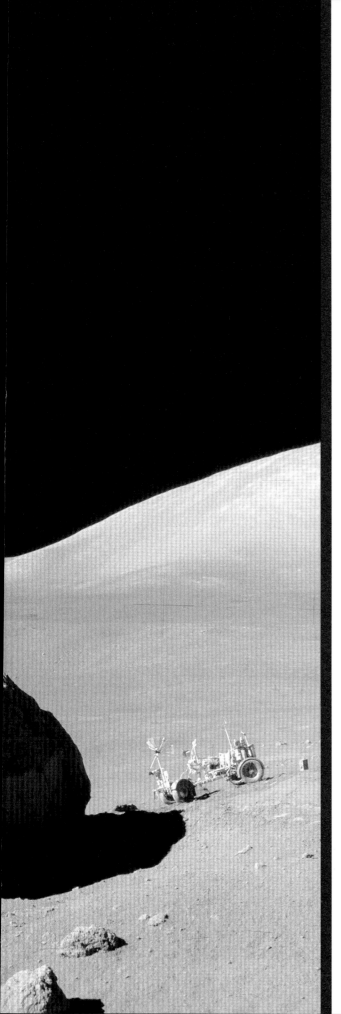

Chapter Three

Electrical and Thermal Control

It is 25 July 1971 and 80m above the launch platform near the top of an Apollo Saturn V space vehicle, engineers manhandle two heavy batteries through an access hatch into the cramped space within. Shrouded by the conical walls of an adapter section is a stowed spacecraft, the Apollo 15 Lunar Module, and in 18 hours it will leave for the mountains of the Moon.

OPPOSITE On the last day that a Rover drove across the Moon, LRV-3 is parked alongside a large split boulder, *aka* Tracy's Rock after Gene Cernan's intention to write his daughter's name in the boulder's dust. Fellow moonwalker Alan Bean eventually made it so in one of his paintings. The LM, over 3km away, is visible just to the right of the boulder's summit. *(Gene Cernan/NASA/David Woods)*

Once inside, they gingerly carry the fully charged batteries around to where a Lunar Roving Vehicle has been folded into a bay on the spacecraft's lower section. There is barely room to move, let alone remove the vehicle's floor panel and shift in these heavy blocks, yet they must be eased in on rods to their almost inaccessible home within the folded vehicle, and firmly bolted in place for the rough ride on the rocket.

Their power connections must all be securely made and their thermal blankets carefully replaced to protect them from the cold of space and the heat of the Sun.

There is no means to monitor the batteries from now on. Until the Rover is deployed onto the surface of the Moon six days later, none of the LRV team, or the crew for that matter, will know if they will have the power to bring the very first Lunar Roving Vehicle to life. It's something that worries all the engineers who built her – but there is nothing they can do about it.

Electrical power

Power to move

In the earliest days of the Lunar Roving Vehicle programme in mid-1969, NASA stipulated that the LRV's source of power would be electricity supplied by batteries. Before then, designers and thinkers had proposed various schemes for how a Rover could be powered. In 1953, Wernher von Braun drew on his experience developing the V-2 missile for the German army during the war, to put forward a hydrogen peroxide power source (see Introduction). He had used the substance to power the rocket's turbopumps by having it decompose into steam over a catalyst.

Conventional internal combustion engines were rarely considered because although hydrocarbon fuels have a very high energy density, they require heavy engines and a separate oxygen supply to work in the Moon's vacuum. Even if they could be made to work, their exhaust would likely contaminate sensitive

RIGHT The Lunokhod, an unmanned, radio controlled rover. Two were successfully landed on the Moon and each spent months slowly exploring its surface. Power came from solar cells exposed by the lid which was opened each lunar dawn. *(TopFoto)*

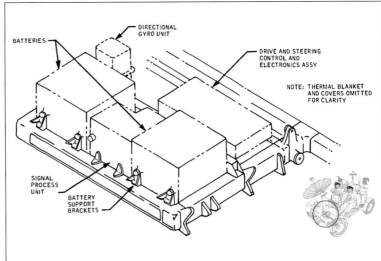

experiments that were trying to sniff out the vanishingly tenuous lunar atmosphere.

Small nuclear power sources are used in spaceflight for long-term, low-power applications. The Mars Science Laboratory, a car-sized unmanned rover launched in 2011, uses a 120W nuclear supply but it is designed to move at slow speeds. One powerful enough for a manned rover would have been far too heavy. Solar cells were employed on Soviet Lunokhod unmanned rovers that could drive slowly over many months but they lacked the power output for the relatively fast sorties planned for an Apollo rover loaded with two suited astronauts and their gear.

In the 1960s, aerospace companies were vying for Apollo work and putting serious thought into how electricity for a rover might be sourced. Two important studies, one by Grumman and Northrop called MOLAB (see Introduction) and another by Boeing and Bendix, both envisaged designs for rovers with pressurised cabins that would have been powered by electricity from fuel cells. Fuel cell technology was then in its infancy and was still being refined through its use in the Gemini space programme. It eventually proved itself during Apollo by reliably powering the Command and Service Module.

Fuel cells work by reacting hydrogen and oxygen over a catalyst to produce electricity and water. But high-pressure storage tanks make these systems too unwieldy for a small, lightweight car that would eventually scoot around the Moon. In mid-1969, when NASA released the specifications for the Rover that they wanted industry to build (see page 34), they stipulated that its source of power would be electricity supplied by batteries. They already had vast experience with this proven technology which was already powering the Lunar Module. It offered reliability, portability and flexibility allied with high power.

Power demands

Nearly every system on the LRV depended on an electrical power supply and so it had to be robust with built-in redundancy. As a result, Boeing dedicated a quarter of the Rover's mass allocation to two 27kg batteries that were mounted on the forward chassis. These would power the four individual wheel motors, the steering motors, the navigation system and, if required, the communications system mounted at the front.

All in all, the Rover would need at least 2kWh for a whole mission and to guard against the situation in which one battery failed, a single battery would have to be capable of supplying all that power. At full load, the electrical system would also have to be able to supply up to 30A at any one time.

FAR LEFT An Apollo fuel cell. Three of these units, each over a metre tall, powered the Apollo Command & Service Module. But, for the rover's needs, they were large, inflexible and the technology was too immature for the conditions. *(Scott Schneeweis)*

ABOVE Layout of the Rover's forward chassis showing placement of the two batteries and the nearby electronics packages. *(NASA)*

UNDERSTANDING ELECTRICAL POWER

Electricity was the lifeblood of the Lunar Rover, as it is for so many of our devices in the modern world. To supply its needs, it carried two large batteries, appropriately selected because running out of juice on the Moon was not an option.

At the heart of understanding electricity are two concepts, voltage and current, and both of these relate to the movement of something around a circuit, a bit like water flowing around a pipe. The exact nature of electricity belongs to the realm of physics, but at a simple level it is a flow of electrons around a circuit and these electrons carry electrical charge. Just as water can be made to turn a turbine, electricity can do work for us.

Continuing the water analogy, we might use a pump to get the fluid moving through the pipe and that pump will provide a force or a pressure. The higher the pressure, the more water is moved around the circuit. A battery provides the force, called **voltage**, that moves electricity around a circuit; the higher the voltage, measured in *volts* (V), the more electricity flows around the circuit.

Water time again – and we can think about how much water passes through a pipe per second. The faster the water flows, the larger the quantity that passes per second and we call the size of this flow its **current**, a term that is also applied to electricity. Measured in *amperes*, or *amps* (A) for short, electrical current is the quantity of electrical charge that is carried by the electrons around the circuit in a given time.

If the push on the electrons (the voltage) changes, then the number of electrons flowing (the current) will also change. By doubling the applied pressure (doubling the voltage) we will double the current. Exactly how much current flows then depends on how easily it can get around the circuit and again, our water analogy comes to our aid. Water flowing through a pipe experiences friction with the pipe walls which slows its flow, and if there are any kinks in the pipe or any other obstruction, these too reduce the current of water flowing by.

In electrical circuits, we describe obstructions to the flow of electrons along a wire using the terms *resistance* and *load*. The greater the load, the less current will flow for an applied voltage. The current will still rise and fall as the voltage rises and falls. It's just that in the case of a large load the current will be less overall although, unlike water, it isn't slowed.

When the current passes through a load, it does some kind of work – turning a wheel motor, running the processors in a smart phone or lighting a bulb. Whatever it is, we are consuming **power** which we measure in *watts* (W). We can work out how much power a load is consuming merely by multiplying the voltage by the current. Thus if a 12V battery maintains a current of 2A through a light bulb, the bulb will be consuming 24W of power.

If we apply the same thinking to the Rover, then when its motors were running at full power, each consumed around 186W. Since they were being fed from 36V batteries, the current to each motor was therefore roughly 5.2A.

What concerns us, and was of particular importance on the Moon, is the question of how long our battery will last; something that is measured in **amp-hours** (Ah). Returning to our 24W light bulb, if we find that our battery is only able to sustain a 2A current through the bulb for 4hr before becoming exhausted, then we can say that the battery had a capacity of 2 × 4 = 8Ah. If the same battery were used to power other appliances, we would find that a lower current would be sustained for longer and a larger current would exhaust the battery faster; for example 1A for 8hr, or 4A for 2hr. In all cases, by multiplying the current and the time we get the same answer (8Ah) and this constitutes the *capacity* of a battery.

For comparison, a small button cell as used in a watch has a capacity of about 0.3Ah, a rechargeable battery for a smart phone has about 1.3Ah, and a laptop battery stores about 5Ah of electricity. At the higher end of the scale, a large car battery will store at least 60Ah. The Rover's batteries were rated at 121Ah each.

Another way to think of the capacity of a battery is to ask how much power it will deliver before it is exhausted. If our 24W light bulb was able to burn for 4hr, that would be the same as a 96W light bulb being supplied for 1hr before exhaustion. This defines the concept of a **watt-hour** (Wh). In this case the battery contained a total of 96Wh of electricity. If we apply this arithmetic to the Rover's two 36-volt, 121Ah batteries, since watts is volts times amps, then it provides

36V × 121Ah = 4,356Wh per battery.

A more common way to express a quantity of power is to divide this answer by 1,000. This is the standard unit for buying and selling electrical energy on Earth. So each Rover battery could provide about 4.35kWh of power, giving the Rover a total power capacity of 8.7kWh.

Another system that required power was the Lunar Communications Relay Unit (LCRU) that was normally attached to the very front of the vehicle. This could be powered from the Rover's batteries but it came with three interchangeable batteries of its own as a contingency in case the Rover were to completely fail. It could then be carried by an astronaut in order to maintain communications with Earth as they walked back to the Lunar Module.

As engineers got to grips with the Rover's actual consumption on Apollo 15, they decided that for subsequent missions, the LCRU would become another load for the LRV batteries, including the TV camera and the transmitters that sent their voices and pictures to Earth. Its dedicated batteries would be still available but their number could be reduced by one.

Batteries

The technology chosen for the LRV batteries used silver oxide-zinc chemistry. The high cost of silver has kept this process on the fringe of battery technology, but on the Apollo spacecraft, where cost was less of a concern, it was used extensively thanks to its high energy density – meaning that a small volume can store a lot of power. It was also tolerant of extreme temperatures and was free of thermal runaway effects. By the time of the LRV project, NASA engineers were already familiar with silver oxide-zinc batteries and with Eagle Picher, the company that supplied them.

Each battery was literally a 'battery of cells' in which 23 silver oxide-zinc cells were connected in series to sum their individual voltages to obtain the desired total of 36V, three times the voltage of a typical lead-oxide car battery of today. With both batteries, the LRV had more than 8kWh of electric power available for its mission. For comparison, a typical household uses 9kWh daily, while for cars, the Toyota Prius hybrid battery has just 1.3kWh available and the high performance all-electric Tesla Roadster sports car has a whopping 56kWh when fully charged. The Rover's total capacity was three or four times more than was ever used on a

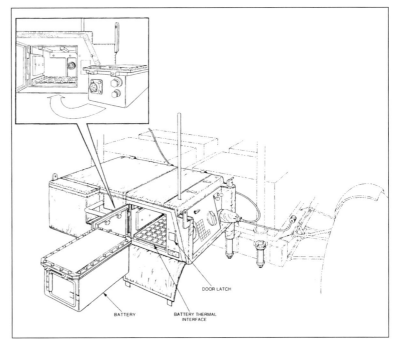

ABOVE The LCRU was mounted on the front of the Rover and had a receptacle for the replacement of its battery. *(NASA)*

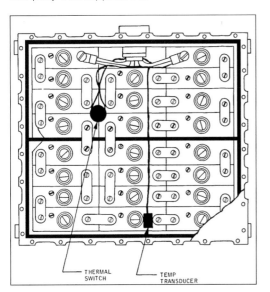

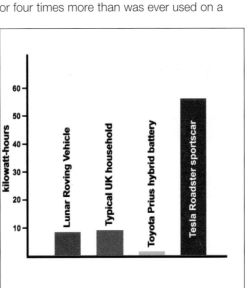

FAR LEFT Layout of cells inside one of the Rover's batteries. *(NASA)*

LEFT Comparison of the power of three battery systems and the consumption of a typical household. *(David Woods)*

RIGHT AND BELOW RIGHT Side and end views of the Rover batteries. Dimensions are in inches. *(NASA)*

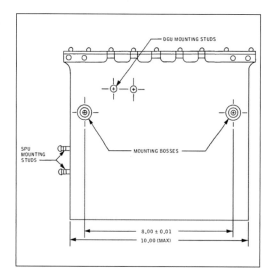

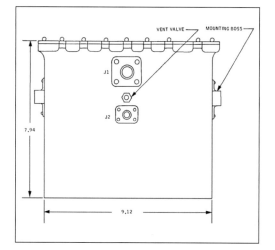

RIGHT Engineers install LRV-1 into a bay on the side of Apollo 15's Lunar Module. The panel that covers the main chassis has been removed to reveal the receptacles for the LRV batteries at the bottom. This is the forward chassis folded in against the central chassis. The white squares at the top are bags placed under the seats where the crew can store items like cameras and rock samples. *(NASA)*

mission but this reflected the policy of having redundant batteries.

> *The 1G Rover drive motors operated from a nominal DC voltage of 34V.*
>
> *The 1G Rover used two rechargeable nickel-cadmium batteries with a capacity of 24Ah each.*

Eighteen hours before launch, the batteries were installed in a fully charged state while the LRV sat folded on the Lunar Module housed within the Saturn V. This proved to be a particularly tricky task because the heavy batteries had to be manhandled into the tight confines of the LM's housing. The process began by removing the panel that covered the underside of the Rover. In the vehicle's folded state, this panel faced to the outside. This allowed access to the folded forward chassis where the batteries would be installed. Once in place, this would be the last time they were checked because there was no monitoring and no recharging took place during the mission.

Each cell in the battery was constructed and sealed in a Perspex case and consisted of two zinc plates connected to the negative terminal which sandwiched a plate of silver oxide connected to the positive terminal. Separator

BELOW Schematic of a single silver oxide-zinc cell. *(NASA)*

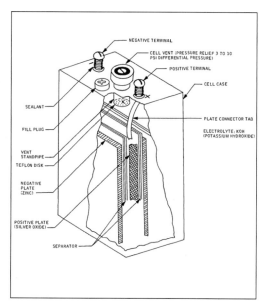

layers kept the plates from physically touching and the whole stack sat in a fluid called an electrolyte, a solution of potassium hydroxide and water. When fully charged, each cell of a battery produced about 1.6V with a capacity of 5Ah, about the same as a laptop battery.

The reaction to produce electricity converted the silver oxide to silver while the zinc became zinc oxide.

$$Ag_2O + Zn \rightarrow ZnO + 2Ag^+ + 2e^-$$

This reaction produced an excess of electrons at the zinc plate and positive ions at the silver oxide plate and it was this difference that enabled the cell to 'pump' electrons around a circuit; in other words, electricity! The process generated some heat which was handled by the Rover's thermal system (see below).

All 23 cells were packed in a lightweight magnesium case and sealed against the vacuum of space to create a single Rover battery. To guard against the water in the cell being electrolysed to produce unwanted hydrogen gas, the case was built to contain a pressure of 9psi and included a valve on one side which opened at between 3.1 and 7psi to relieve any overpressure.

Both batteries were mounted on the forward chassis, in front of the astronauts' feet, along with some of the vehicle's electronic systems; an arrangement that made an unloaded Rover a little front heavy. In normal operation, they shared the electrical load by having the front wheels powered from one battery and the rear wheels powered from the other. The loads from the two steering systems, front and rear, were distributed in the same manner.

Switches on the Control and Display panel enabled the astronauts to change which load was supplied by which battery. In the extreme case where a battery had failed, it was possible for an astronaut to switch all loads to the other battery, but if he did so he had to manage the greater heat generated by the faster discharge. (See section below on thermal management.)

Battery monitoring

While their lives did not exactly depend on the batteries, for they could always walk back to the LM, the astronauts were nevertheless very concerned with their vehicle's health. A long walk would be risky and a failed Rover would

jeopardise the mission's goals. The LRV's Control and Display panel (see Chapter 5) included a set of three meters to enable the crew to keep an eye on the state of their power supply.

The first indicated how much charge remained in the batteries, a quantity normally given in Ah. This is a bit like a fuel gauge in a petrol or diesel car. The easiest way to think about an amp-hour is to imagine a battery with

LEFT The forward chassis of LRV-3 during the early moments of Apollo 17's lunar exploration before dust had covered everything. The batteries and electronics are protected within white thermal blankets and the covers are closed. *(NASA)*

BELOW Schematic of the Rover's Control and Display panel. Dimensions are in inches. *(NASA)*

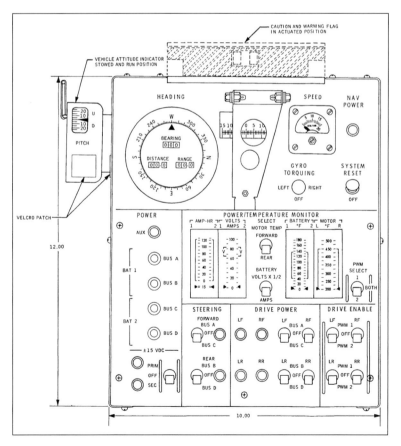

RIGHT This set of meters, seen here on LRV-2's Control and Display panel, showed the state of the batteries and the temperatures of the motors. *(NASA)*

a capacity of 1Ah; a typical phone battery. This could supply a current of 1A for 1hr, 2A for 30 min or a 0.5A for 5hr. For the Rover batteries, each had a rated capacity of 121Ah of which 105Ah was considered useful. For comparison, the capacity of a typical lead acid battery in a car is 45–60Ah.

Next to the amp-hour meter was one which could display either voltage or current, depending on the position of a switch. This meter was scaled 0 to 100 and if switched to voltage, an astronaut had to divide the reading by two to get the correct value.

Finally, a thermistor in each of the battery cases provided the signals for a pair of temperature gauges. The range of this meter (calibrated in Fahrenheit) included markings to indicate the acceptable range of temperatures for the batteries of 40–135°F (4.5–57°C). While the latter two meters were fairly conventional, the meter that showed remaining charge got its information from an unusual device.

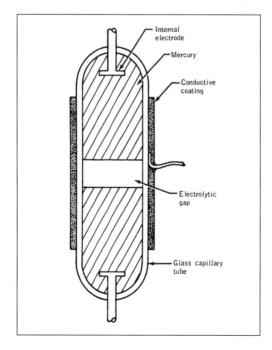

RIGHT Schematic of the mercury coulometer, the device at the heart of the Rover's amp-hour meter. *(NASA)*

A meter for electric charge has to count how much has passed and for this, engineers used a very thin glass tube which was mostly, but not completely filled with mercury. A small gap in the mercury split it into two columns, one above the other, and that gap was filled with a liquid that acted as an electrolyte. As the batteries discharged, a little of their current was shunted through the contents of this tube. Where it crossed the gap, it caused the mercury atoms to migrate across the electrolyte from one column to the other. As charge continued to pass, one column got shorter as the other lengthened and so the length of these columns changed in proportion to the amount of electric charge that had passed. Such is the ingenuity of engineers.

To drive the charge meter on the Control and Display panel, the position of the gap had to be measured and to do this, they turned the tube into a capacitor. At its simplest, a capacitor is a device that stores electric charge across two conducting plates. Between the plates there is some form of insulator, be it vacuum, air, glass or some other material. In this case, the wall of the glass tube was the insulator.

To turn it into a capacitor, the outside of the glass tube was coated with a conductor and this formed one side of the capacitor. The other side was formed by one of the mercury columns. As the length of the column changed, so did the ability of the capacitor to store charge, its so-called 'capacitance value'. A simple bit of electronics could record this capacitance to generate a voltage that could drive a meter. In practice, these battery charge meters proved to be somewhat erratic, and astronauts were known to tap the panel to try and get a consistent reading from them.

There was one other part to the Rover's monitoring system, a caution and warning flag. At the top of the Control and Display panel, fixed to its housing, was a hinged strip, the flag, that was spring-loaded to stand erect and display a gold-coloured striped pattern to the astronauts. Normally it lay folded against the housing, held in place by a magnet. But if any of the Rover's thermal switches closed due to excessive temperature in the batteries or in one of the motors, a short 36V pulse would be sent to an electromagnet to release the flag and let it pop up in the astronaut's field of view (see Chapter 5).

The trigger temperatures for these switches were 52°C (125°F) for the batteries and 204°C (400°F) for the motors. Once the flag had been raised, the astronaut was able to push it back down to reset it.

Power conversion

Not all the devices on the LRV could work directly off the 36V direct current (DC) supply from the batteries and so both the Signal Processing Unit (SPU) and the Drive Control Electronics (DCE) included extra circuitry that acted as power supplies as necessary.

The navigation system (see Chapter 4) included a directional gyroscope that needed a 115V, 400Hz alternating current (AC) supply. This was generated by an inverter in the SPU along with DC supplies of 5V, 28V as well as +/–15V for the unit's electronics. Likewise, the electronics for the T-handle and the DCE required +/–15V DC supplies which were generated by a circuit board contained within the unit.

Thermal control

On our home planet, we are all instinctively aware of how heat and cold change around us; the cosy ambience of our homes against the cold outside, the daily change

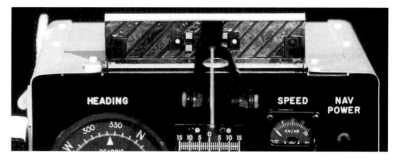

ABOVE The caution and warning flag on LRV-2, shown here in its actuated state. *(NASA)*

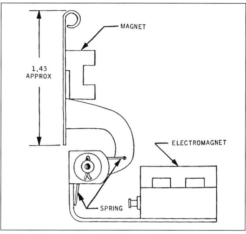

LEFT The mechanism of the caution and warning flag, shown in its actuated position. *(NASA)*

BELOW The shots from which this panorama was made were taken by John Young towards the end of Apollo 16's second day of exploration. Charlie Duke is working at LRV-2. The LCRU is nearest the camera with its thermal blankets pulled back to reveal its mirrored top surface that radiated heat away into space. Behind is the forward chassis with its batteries and electronics, all of which were protected by thermal blankets which were dust-covered by this time. *(John Young/NASA/David Woods)*

between night's chill and the day's warmth, the annual cycle of the seasons. All around us, temperatures are influenced and moderated by our environment as the atmosphere and the sea constantly transport heat from places warmed by the Sun to those that are shielded from its rays.

Although the Moon is as far from the Sun as we are, and receives the same intensity of light and heat from it, the lunar surface is a very different place. It possesses neither an ocean nor an atmosphere to cushion the extremes of temperature. An object placed on the Moon during the day will quickly get hot on its sunlit side whilst simultaneously, its shaded face will become profoundly cold as it radiates its heat to deep space.

There is no rising convection current of warmed air to take heat away from a hot surface. Instead, as soon as the Sun rises over a lunar landscape, as it does each month, a rock that sits face-on to direct sunlight will heat to around 125°C. At the same time, with no blanket of air to warm it, its unlit face will remain extremely cold and struggle to rise above −150°C.

To add to these thermal problems, the astronauts intended to stay for three days and during that time the Sun was going to get much higher in the sky. On Earth, daylight lasts for an average of 12 hours, but on the Moon, there are 360 hours between sunrise and sunset, so that by lunar noon the entire landscape has stabilised at 125°C. For comparison, because things cool down during the fortnight-long lunar night, the pre-dawn temperature can dip below −180°C.

Planners deliberately chose to land in the early lunar morning, when the Sun had brought the average surface temperature up to around 20°C. This kept thermal loads moderate for the early Apollo landings that only stayed for a day or two. But by the end of a stay lasting three Earth days, the temperature had risen to 70°C. By then, the heat was coming not just from the Sun. The entire landscape was radiating thermal energy that bathed the astronauts and the Rover.

The Rover's mix of batteries, electronics and mechanical devices had to be protected from these extremes of heat and cold, for they could only operate or even survive across a narrow range of temperatures.

It was thermal engineers Hugh Campbell's and Ron Creel's job to ensure that all the Rover's systems and components were maintained within their temperature limits. "You've got to be able to get rid of the heat when you need to get rid of the heat and store it when you need to store it, keeping the temperatures of your components within a fairly tight band of temperature requirements," he explains.

The batteries were the most temperature-sensitive items, and in order for them to work they had to be maintained in the range 4.5–57°C. When not in use, they could not be allowed to go colder than −26°C or hotter

ABOVE AND RIGHT In these drawings of the Rover, those areas that received special finishes for the purpose of thermal control are marked in black. *(NASA)*

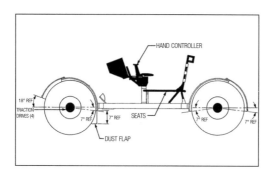

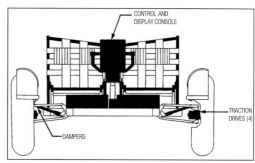

than 60°C because of the water-based electrolyte they contained. And even though the most delicate mechanism of them all, the astronaut, was protected by his space suit, it was important to ensure that none of the LRV's surfaces he might touch could become so hot that they could damage the suit.

In addition to protecting against these extremes of temperature, the thermal system had to deal with the fact that many of the Rover's electrical systems also generated their own heat, and this needed to be carefully managed. In a normal car, liquid coolant is used to control the temperature around the engine. A mixture of water and glycol takes heat from where it is not wanted, like the engine block, puts a portion of it where it is wanted, like the car's interior on a cold day, and rejects the rest to the atmosphere via the radiator. The two Apollo spacecraft, namely the Command and Service Module and the Lunar Module, used very similar methods employing pipes, pumps and a water/glycol coolant to transport heat. The Service Module even had large radiators to get rid of excess heat.

For the Rover, engineers were faced with the fact that a conventional thermal system would be far too heavy. Creel points out that their mass budget for solving all the thermal management problems was set at a mere 4.5kg! "We counted ounces and tenths of ounces during the development programme," he grins.

With such tight constraints, their solutions typically employed a range of techniques, often ingenious, to maintain control of the temperatures. Some were simple and passive, like painting surfaces white or making sure that the Rover was never parked in the shade of the LM where it would become too cold. Others were a little more elaborate, like using the thermal capacity of the batteries to moderate the temperature of nearby electronics.

There were no parking limitations for the 1G Rover.

Battery temperature

Management of battery temperature was a major issue because they generated a lot of heat as they discharged during a drive. This heat was allowed to build up in the batteries by taking advantage of the fact that the electrolyte in the cells contained water. An important property of water is its huge capacity to store heat energy. For example, it takes over four times as much energy to raise the temperature of a gram of water by one degree than it does for a gram of aluminium.

But even with such a great heat sink available, at some point the batteries would produce more heat energy than they could contain; and getting rid of this excess heat was difficult because each battery was in a cocoon of thermal blankets to shield it from the heat of the Sun and the chill of deep space. These blankets were made from 15 layers of aluminised Mylar, a thin plastic film which was made reflective in order to prevent heat passing through as infra-red radiation. Between each layer was a nylon net that kept them apart. The outer layer of the insulation was made from white fire-resistant beta cloth, a material made from glass fibres coated with Teflon which was also used for the outer layer of the space suits. To further isolate the batteries from the external temperature, they were mounted

BELOW Diagram to indicate the preferred parking attitudes with respect to the Sun. It shows that when the Sun got higher in the sky, the Rover should be parked with its right side facing the Sun. *(NASA)*

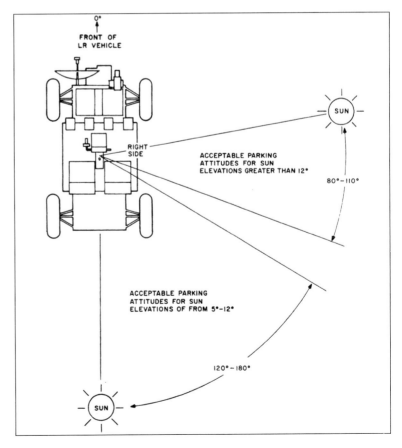

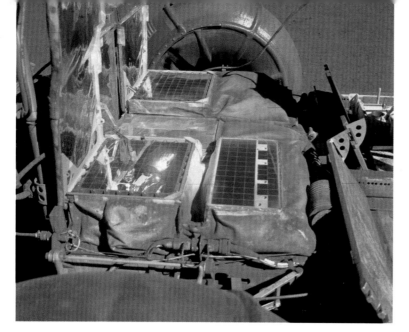

ABOVE The forward chassis of LRV-1 after the vehicle had been parked for the last time during the Apollo 15 mission. The covers are raised to reveal the mirror-like radiators at the top of each battery. The white thermal blankets that surround each battery have been largely covered with lunar dust. Battery 1 is nearest the camera with the DCE radiator to its right. Battery 2 is beyond. *(David Scott/NASA)*

BELOW Top and side drawings of the LRV battery and its thermal support including dust cover assembly, fibreglass mounts and thermal straps. Dimensions are in inches. *(NASA)*

using fibreglass brackets that minimised the conduction of heat to and from the chassis.

Cooling the batteries was a job that the astronauts would have to help with. Occasionally, when the LRV came to a stop, one of their tasks was to open the battery covers. This exposed a mirrored plate attached to the top of each battery that acted as an efficient radiator to space. At the same time, the mirrors were able to reject the Sun's heat by reflecting it away, and this worked best when they were clean. During a drive, the covers were kept closed to protect the radiator from the ubiquitous lunar dust that otherwise would spoil its cooling properties.

Once the covers were open, the problem was to stop the batteries getting too cold. "What if the crew didn't come back to the Rover for 14 or 15 hours during a rest period?" asks Creel. If the electrolyte froze, the expanding ice might crack the battery case and destroy it.

The thermal engineering team's solution was to hold the covers open with a latch. Then, when the batteries got down to around 5°C, a

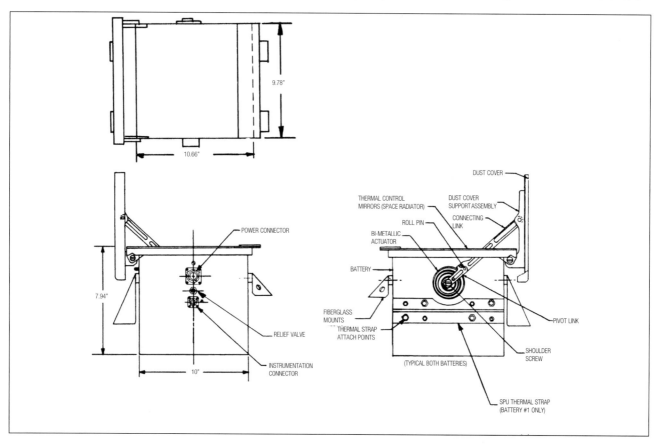

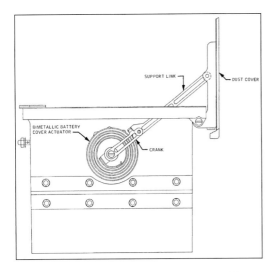

coiled bimetallic spring reacted to the cold by disengaging the latch to allow the covers to close. "They were similar to a thermostat you have inside the water heating-cooling system on cars," says Creel.

The bimetallic spring consisted of two long strips, one made of copper and the other of manganese. These were bonded together to form a single strip and shaped into a coil. When these metals warm, manganese expands a little more than copper and this difference in expansion made the strip change its amount of coil and thereby produce a turning force that unlatched the mechanism and closed the covers.

This somewhat 'Heath-Robinson', mechanical closing mechanism was one of the astronauts' favourite parts of the Rover, and a nice complement to its more exotic, sophisticated electronic systems.

Electronics thermal control

Electronic systems are a little more tolerant of temperature than liquid-filled batteries. Nevertheless, engineers had to take care of the Rover's electronics because temperature extremes will affect them too. Different materials show different degrees of thermal expansion (as indicated above), so where two materials are fastened together they might come apart as they expand and contract at different rates. For electronics, printed circuit boards are particularly susceptible because the copper tracks on their surface can crack under the stresses of thermal expansion, which then breaks the circuit. Other electronic components such as capacitors and transistors are also temperature sensitive.

Three of the systems which needed thermal regulation were mounted beside the batteries on the forward chassis: the Drive Control Electronics (DCE) that fed power to the motors; the Signal Processing Unit (SPU) that performed all the navigation calculations; and the directional gyroscope which although it was not strictly an electronics system, had a similar need to lose heat.

A fourth electronics unit that needed careful thermal handling was the Lunar Communications Relay Unit (LCRU) which was attached by the astronauts after deployment and stuck out of the front of the vehicle. It was very much a separate system and had its own thermal protection.

Like the batteries, the DCE, SPU and gyroscope units were mounted to the chassis by brackets made from fibreglass to limit conduction to and from the chassis structure. The units were also protected from their surroundings by the same thermal blankets and insulating dust covers that protected the batteries, so they required a way to lose their own internally generated heat. Mirror radiators were again used to dissipate excess heat and this was also true of the LCRU, but the engineers had other tricks up their sleeves.

Engineers made use of the considerable ability of the batteries to absorb heat by bolting the SPU and gyroscope to the battery cases using aluminium brackets called thermal straps. This allowed heat to flow into the batteries, from where it could later be released.

ABOVE LEFT The bimetallic strip mechanism that automatically closed the protective covers over the batteries when they had cooled sufficiently. *(NASA)*

ABOVE Thermal arrangements for the Drive Control Electronics and battery 1, including fibreglass mounts, radiators and wax tanks (fusible mass). *(NASA)*

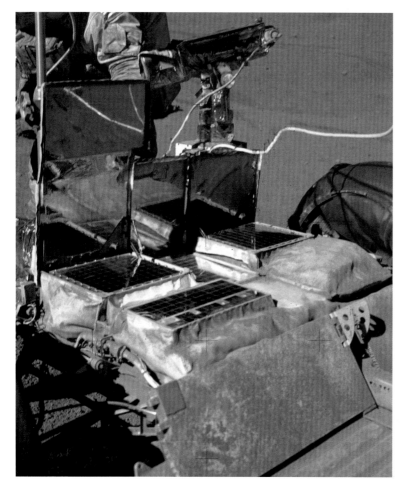

ABOVE The forward chassis of LRV-2 after it was parked in its final resting position. The covers are open to expose the mirror radiators to space and let the batteries cool. There is much dust on the vehicle after one of the fenders got damaged. *(John Young/NASA)*

BELOW Thermal arrangements for the forward chassis, including fibreglass mounts, radiators and wax tanks (fusible mass). *(NASA)*

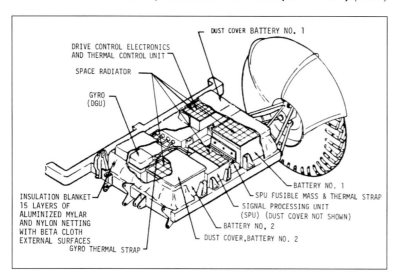

During long drives, when the mirror radiators were covered up, all three electronics units – the LCRU, the SPU and the DCE – employed an unusual means of thermal control that made use of the thermal properties of tanks filled with paraffin wax, particularly the 'phase change' from solid into liquid. As heat is applied to a substance, the energy goes into making its molecules vibrate faster and as a result, its temperature rises. For certain solid substances, there will come a point when the vibration of the molecules begins to overcome the bonds that hold the material together and it will begin to melt. At the melting point, adding more heat does not raise the temperature further. Instead, it causes more material to melt to a fluid and so the rise in temperature is stalled until it is completely melted. The heat consumed in this phase change is known as latent heat.

Engineers put three tanks of wax in the LCRU and one each on the DCE and SPU. Wax has a very high specific heat capacity, almost as good as water, and combined with its phase change at about 37°C, it could deal with a lot of the excess heat from the electronics. By absorbing heat as it melted, it would limit the rise in temperature of the electronics. When the covers were opened and the mirrored plate radiators exposed to the chill of space, the wax also gave up its heat and resolidified, ready for the next cycle.

To ensure a good contact between the wax tank and the electronics case that it was attached to, Ron Creel and his colleagues at Marshall had suggested trying a rubberised bonding RTV material. After a routine test at Boeing, in Seattle, one of the thermal engineers had been asked to find out how good the contact had been. Creel takes up the story. "That fellow interpreted that, to go back and prise it off. Well in putting the crowbar on it and prising it off, he damaged the surface of the DCE box that was underneath. I came in the next morning and everyone was running around like they were going to fire this poor fellow because of what he had done. I had to call back here to Huntsville, to start up through the chain of command there and get the fellow's job saved. He was just trying to do something to help us out and we didn't want to get him in trouble!"

Control and Display panel

Another part of the LRV that needed special attention from the thermal engineers was the unit that housed the instruments and controls. It sat high above the chassis at the crew's eye line. It contained a few delicate electronics, which along with its meters and its heading display, were sensitive to extreme cold. The main tactic to protect it was to ensure that at the end of the astronauts' day, they remembered not to park the Rover in the shadow of the Lunar Module where temperatures fell well below −100°C.

The unit was housed in a box that was painted white in order to reflect sunlight and inhibit heat loss by radiation. However, the Control and Display panel itself had to be black to minimise glare into the astronauts' eyes, and a black surface will heat and cool much more readily than a white one. Knowing that a black panel would get very hot in the Sun, engineers mounted it on stand-offs or 'feet', which separated it from the structural panel below. The low thermal conductance of these feet reduced the heat flow between the panel and the rest of the unit.

Dust control

Much of the Rover was white for very good reason. When compared to a dark surface, white absorbs less infra-red heat, reflecting most of it back into the environment. It also emits less heat by infra-red radiation, which keeps a warm object warmer for longer. But when the Rovers got into the field, the engineers' careful schemes to control temperature by reflecting heat were nearly undone by one phenomenon – the lunar dust.

"We had done testing here on Earth that had shown that dust on radiator surfaces, as well as other thermal control surfaces would act as an insulator," says Creel, "preventing the radiators being able to get rid of the heat."

This was an unfortunate problem for a vehicle designed for the Moon. Most of the lunar surface is profoundly ancient. Much of it solidified well over 3 billion years ago, and since then only one thing has ever really made any difference. Over the aeons, without an atmosphere to protect it, an incessant rain of hypervelocity meteoroids has bombarded the

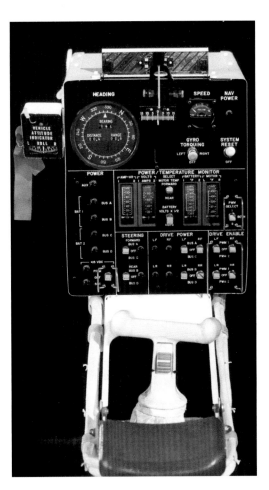

LEFT LRV-2's Control and Display panel prior to flight. *(NASA)*

BELOW Harrison Schmitt at work at Camelot Crater near the end of Apollo 17's exploration of the Taurus-Littrow valley. Much of his suit is covered in dust. *(Gene Cernan/NASA)*

surface, gradually pulverising the exposed rock into finer and finer dust. This bombardment has created the regolith, a layer typically 5m deep that blankets almost all the Moon.

As a result, most of the regolith exposed at the surface consists of an extremely fine dark powder of jagged rock fragments which are both extremely abrasive and very good at absorbing the Sun's heat. Moreover, in the super-dry environment of the Moon's vacuum, the static electric charges that build upon anything that moves makes these dust particles readily cling to items taken there by humans. It was not uncommon for an astronaut to fall over, and when he got up he was often as filthy as a coal miner.

For the Rover's designers, the plan was to keep dust off the vehicle as much as possible and shield mechanical components from its ingress. The motors and transmission units were built into sealed cases. Silicone rubber 'boots' were installed where levers had to enter systems, such as where the brake levers entered the brake drums and where the T-handle emerged from the box that held its mechanism.

The main defence against dust was the fenders around the wheels (see Chapter 2) which countered the property of the wheel mesh to lift dust and throw it into 'rooster tails' behind as the Rover travelled forward. An intact set of fenders was able to handle these wheel dust sprays well, but on all three Rover missions, the fragile fenders proved to be no match for a busy astronaut in a cumbersome suit whose helmet provided limited visibility.

After Gene Cernan accidentally ripped off part of a fender at the start of LRV-3's time on the Moon (see Chapter 7), he and the engineers in Houston were well aware that the Rover would be showered with dust. A quick attempt to fix the fender with duct tape failed as the tape could not adhere properly to the dusty surfaces, so overnight, engineers came up with a successful fix that used discarded stiff maps, duct tape and clamps normally used to hold small lights in the Lunar Module. (See Chapter 7 for the full procedure for this repair.)

BELOW As Apollo 16's exploration of the Descartes landing site came to an end, LRV-2 was parked up for the final time. Here, Charlie Duke has just finished dusting the battery radiators, having opened their covers. His suit is very grimy from 20 hours' hard work on the lunar surface. *(John Young/ NASA)*

This ingenious repair to LRV-3's fender was a triumph that echoed similar virtuoso solutions from the flight of Apollo 13. However, using a large brush that had been thoughtfully included in their kit, the astronauts had a constant battle to keep the surface of their Rover and themselves clean, especially the radiators that were to maintain its systems within acceptable temperature limits. As a result, LRV-3 experienced higher thermal loads than its predecessors and the crew spent a significant time repeatedly brushing down the vehicle, its TV camera and each other!

The fact that a little lunar dust could so easily push up the Rover's finely balanced temperature illustrates just how tight the engineering margins were when it came to building a car for the Moon. Perhaps the greatest tribute to those who designed and built the vehicle is that in a total of over 50 hours spent driving across 90km of rugged lunar terrain, none of the three LRVs ever let their crews down, and only required a bit of duct tape and a few discarded geology maps to keep them running!

ABOVE A detail from a photo of the front end of LRV-3 shows the brush used to keep the Rover dusted. The exposed LCRU radiator is on the right. *(NASA)*

BELOW LRV-3 with its makeshift fender towards the end of Apollo 17's second day of exploration in a panorama created from images taken by Jack Schmitt. The Rover is at a geology stop known as Station 5, where a scattering of boulders mark the rim of the 700m crater Camelot.
(Harrison Scmitt/NASA/David Woods)

Chapter Four

Navigation and Communication

It is 6 February 1971 and two Apollo 14 astronauts are experiencing the difficulties of navigating on the surface of the Moon. Alan Shepard and Edgar Mitchell are the first men to roam so far on foot from the relative safety of the Lunar Module. About 1km from their landing site and approaching the limits of their air, they are still searching for the rim of Cone Crater; the principal scientific target of their mission. And Mitchell is beginning to think they might never find it.

OPPOSITE Fra Mauro: the landing site for the Apollo 14 Lunar Module, Antares. It is a bleak, desolate landscape with few landmarks, photographed here by astronaut Alan Shepard and formed into a panorama. Navigating this wilderness will be tricky without a global magnetic field to aid them. In the middle distance Ed Mitchell sets up a colour TV camera to send coverage of their exploration to Earth. But once aimed, it cannot move and it will fail to cover their ascent up the flank of a nearby ridge to inspect Cone Crater. *(Alan Shepard/NASA/David Woods)*

"It's going to take longer than we expected," he radios to Fred Haise in Mission Control. "Our positions are all in doubt now, Fredo." The undulating terrain of Fra Mauro has disorientated them. Out of sight of the TV camera, which was on a tripod at their landing site, both Apollo 14 astronauts and the teams at Mission Control have lost track of their position. In Houston, thoughts are turning to telling the two men on the Moon to turn back.

"Okay, Al and Ed," replies Haise. "The word from the Backroom is they'd like you to consider where you are [to be] the edge of Cone Crater." They weren't going to go any further on this trip.

"I think you're finks!" retorts Mitchell, frustrated that the two of them would have to turn back before reaching a destination that would have given them a spectacular view across a hole in the ground some 340m in diameter.

The LRV engineers at NASA, Boeing and General Motors, listening on the air-to-ground loop around the country, are only too aware that in just five months' time the next Apollo crew will be driving the first Lunar Rover over ten times further away from the safety of the Lunar Module. Would they get lost too?

NASA had dismissed Boeing's original plans for a full inertial guidance system, similar to that used on the Apollo spacecraft, for being too complicated and too heavy. But would their new, simpler navigation system be up to the job of keeping the astronauts safe? And as if this is not demanding enough, the first Lunar Roving Vehicle will also be called upon to act as an independent interplanetary outside broadcast unit; relaying live colour TV pictures from the Moon and all the conversations between the crew and Earth. The world is relying on the Rover team to provide unprecedented coverage of these final human adventures in the mountains of the Moon.

Navigation

In the pre-sat nav/smartphone world of the 1960s, most navigation that involved wheeled vehicles on Earth was by reading maps and remaining on a road network. Off-road navigation usually relied on a magnetic compass combined with dead reckoning whereby position is calculated only by using heading, speed and time from a previous point

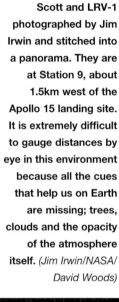

BELOW David Scott and LRV-1 photographed by Jim Irwin and stitched into a panorama. They are at Station 9, about 1.5km west of the Apollo 15 landing site. It is extremely difficult to gauge distances by eye in this environment because all the cues that help us on Earth are missing; trees, clouds and the opacity of the atmosphere itself. *(Jim Irwin/NASA/David Woods)*

to a new one. But a magnetic compass will not work on the Moon because there is no global magnetic field. In addition, there are few useful landmarks and distances are difficult to judge. Yet scientists wanted the Apollo crews to venture as much as 8km across hummocky, hilly terrain to accurately reach sites of scientific interest that were likely out of sight of the LM. Good surface navigation was a necessity if the Rover was to make good on its promise of increasing Apollo's scientific return before the end of the programme.

An early proposal, borrowed from the guidance and navigation techniques that were already working in both of the Apollo spacecraft, would have used a set of three gyroscopes to measure rotation and accelerometers in order to measure movement. NASA, and in particular the astronauts, were unhappy with this proposal because not only was the method untested in an off-road car, it was complicated and costly, and they doubted whether it could be operational in time.

Astronauts are pilots, and pilots were trained to navigate by dead reckoning. If a system could present heading and distance information to an astronaut, that would suffice for them. With the consent of all concerned parties, a much simpler and more rugged navigation tool was developed for the LRV that worked in a way that made sense to pilots. It measured the rotation of the wheels and the direction of travel and calculated the numbers required to guide the astronauts to their next assigned geology stop, and ultimately back to their LM home.

The final navigation system had three elements: a set of odometers, which were really part of the motor assembly, a directional gyroscope and the Signal Processing Unit (SPU) that did a little mathematics. Its results were supplied to a dedicated section at the top of the Control and Display panel. This showed the Rover's heading, its speed and trip distance, and for purposes of navigation it always showed the range and bearing to the LM for use in the event that the astronauts decided to drive straight 'home'.

ABOVE This panoramic view south across the Taurus-Littrow valley was produced from images taken by Gene Cernan on his last day on the Moon. On the right is the South Massif which they had visited the previous day. It is over 10km away and rises 2,200m above the valley floor.
(Gene Cernan NASA/ David Woods)

BELOW The major components of the LRV navigation system. *(NASA)*

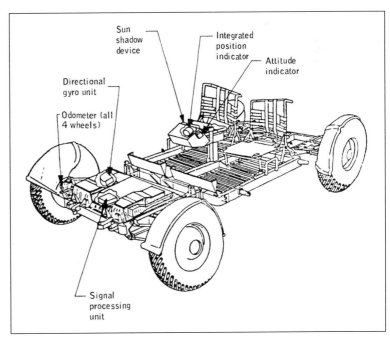

RIGHT This photograph of Gene Cernan was taken at the end of Apollo 17's exploration. It gives a clear view into the left front wheel hub and the cable that carries pulses from the odometer. *(Harrison Schmitt/NASA)*

Odometers and the speedometer

Each wheel's motor unit included an odometer to measure the rotation of the wheel and thereby enable the distance travelled to be calculated. Nine small magnets were arranged around the output end of the unit and a stationary reed switch was attached nearby. The reed switch operated each time a magnet passed by, so that as the wheel turned the switch generated a series of pulses that metered the Rover's progress, nine pulses per rotation.

Not strictly part of the navigation system but related to it was the vehicle's speedometer, which was displayed at the top right of the Control and Display panel. The speedometer's reading was derived from odometer pulses taken from the right rear wheel only. Since speed was not a crucial measurement for navigation, no attempt was made to provide redundancy and there was no provision to take pulses from any other wheel.

It was possible that the pulses from the odometer would be of inconsistent length. But given the way the speedometer worked, these pulses had to be of equal length to be accurate and they were electronically tidied up. Next they were filtered to produce a DC output voltage. A high speed meant lots of pulses and therefore a higher voltage; few pulses produced a low voltage. This voltage was displayed on a meter but rather than showing volts, it was calibrated in kilometres per hour and therefore became a speedometer. Like a car speedo, its full scale was somewhat higher than it was ever going to travel. The scale went to 20kmh, but even on a downhill slope the Rover never actually exceeded 17kmh.

BELOW Diagram of the wheel hub and motor/transmission unit showing the connection for the odometer cable. *(NASA)*

BELOW LEFT The navigation section of LRV-2's Control and Display panel with the speedometer on the right. *(NASA)*

BELOW Few odometer pulses yield a low voltage for the speedometer display. More pulses raise the voltage and move the needle of the meter. *(David Woods)*

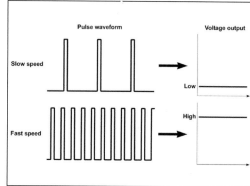

Directional gyroscope

The directional gyroscope used on the Rover was contained in a box 11cm square and 14cm high which was mounted on the forward chassis panel behind battery No 2. This was a standard avionics unit; model 9010 built by Lear Siegler Inc., one of the many US companies that sprang up in the late 20th century to take advantage of increased spending on aerospace and the need for quality flight instruments. Directional gyroscopes such as this were used for heading displays found on most aircraft instrument panels at the time.

The unit contained a single spinning gyroscope mechanism, a mount that allowed that mechanism to rotate, and a device called a synchro transmitter to measure the angle of rotation. The gyro required a 400Hz AC power supply at 115V, and this was generated from the direct current of the batteries by an inverter in the Signal Processing Unit. Once the gyro's rotor was up to speed, the unit consumed about 15W and engineers ingeniously used No 2 battery to absorb the excess heat it produced by physically attaching it to the battery's case (see Chapter 3).

The gyroscope mechanism at its core worked in exactly the same way as a toy gyroscope by having a rotor that was spun at high speed. This made its axis strongly resistant to turning; it always tried to keep pointed in the same direction. If it was aligned to point north, then it would tend to keep pointing north as long as it had freedom to turn, no matter how the unit's case (and therefore the LRV) was orientated. The mechanical gyro was never a perfect device and it was specified that the Rover's should drift away from its alignment at a rate no faster than 10°/hr. In fact, all of the gyros used on the Moon showed negligible rates of drift.

It was the job of the gyro's synchro transmitter to measure the angle between the rotor's axis and the case. This had three

> On Earth, we have been very lucky that our planet is surrounded by its own magnetic field. Not only does it shield us from some of the nastier phenomena that space throws at us, like solar storms, it has given us a useful reference as to which way is north. In truth, we have two north poles; one is the true north that is defined by the planet's axis of rotation and the other is the magnetic north which slowly wanders and is usually a few degrees away from true north.
>
> When the Moon formed out of a great impact that afflicted the young Earth, it failed to gather much iron from the debris and so lacks a substantial iron core that might have given it a global magnetic field. As a result the Moon offers no simple means of determining direction, even though it has a geographical pole around which it rotates. The Sun could be used in this role but it would not work well on a moving car.
>
> A properly engineered gyroscope can act as a substitute for a magnetic compass because, over a limited time and distance, it will remain aligned in a constant direction to which we can refer. In other words, if we align its axis with true north (see Nav Initialisation below), it will tend to maintain that alignment over the course of a few hours. To achieve that alignment we can use one easily visible reference, namely the Sun.

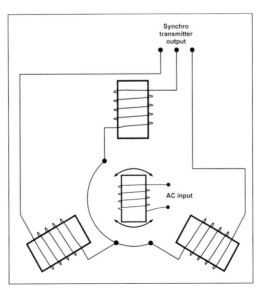

FAR LEFT When the rotor of a toy gyroscope is spun at high speed, it will balance on the end of a pen because it strongly resists changes in the angle of its axis. *(David Woods)*

LEFT Schematic of a synchro transmitter. Apply an AC voltage to the central coil and the output coils will give AC outputs that represent the angle of the central coil. *(David Woods)*

RIGHT The Integrated Position Indicator (IPI) at the top left of LRV-2's Control and Display panel. It includes numeric displays for distance, range and bearing, as well as a compass rose, a circular display of the Rover's heading. *(NASA)*

electrical coils that were fixed to the case and arranged in a circle at 120° intervals. In the middle was another coil that was fixed to the gyroscope's mechanism, which meant it remained aligned to north. Since it was fed with an AC electric current, it created a varying magnetic field that the three outer coils (which rotated with the vehicle) could pick up; with the magnitude of each outer coil's output being proportional to its angle with respect to the central coil. Consequently, the output of these three coils was an electrical representation of the LRV's heading, and their signals were fed to the rest of the navigation systems along three wires that led from the unit.

The signals from the gyroscope unit went to two places on the Rover. One went directly to the Control and Display panel to drive a compass rose display which formed part of what was called the Integrated Position Indicator (IPI). This was a circular indicator like those found in aircraft of the time, and it was marked with north, south, east and west. When the Rover was steered one way, this display rotated in the opposite direction so that it always showed their current heading. The compass rose was driven by a so-called 'synchro follower' device, which was essentially the reverse of the synchro transmitter described above. In other words, it converted the signals on the gyro's three wires into a relative rotation. A second feed of the gyro's output went to the SPU for processing.

Signal Processing Unit

In the centre of the compass rose were three numeric displays labelled 'Distance', 'Bearing' and 'Range'. The role of the SPU was to take the signals from the gyro and the odometers and process them in order to drive these displays. The distance display was basically a trip counter, very similar to those used in cars. The other two displays told the astronaut how far away the Lunar Module was as the crow flies (not that there were any crows on the Moon and even if there were they wouldn't be able to fly because there was no atmosphere) and what bearing he should take to return to it.

Although often characterised as a computer,

BELOW Astronaut Bob Parker works with a 16mm movie camera on the qualification version of the LRV. In front of him is the Control and Display panel with markings that show it is not intended for flight. *(NASA)*

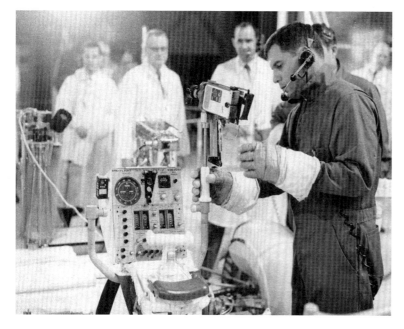

BELOW Diagram of the Integrated Position Indicator (IPI). *(NASA)*

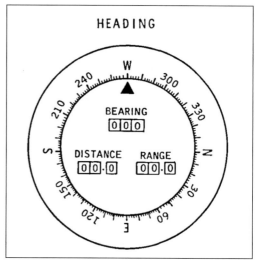

this Boeing-built unit was a simpler and rather specialised piece of equipment for that term, at least in the modern sense. It used a mixture of analogue and digital techniques to do its work and it was not programmable. Its own power consumption was 28W, and the excess heat was absorbed by battery No 1, to which it was attached (see Chapter 3). It acted as a power supply for the gyro and compass rose by converting the 36V DC supply from the batteries to the AC voltages they required.

The distance display was the least complex of its main tasks. First, the SPU counted the odometer pulses coming from all four wheels, three pulses at a time so as to gain an average. Its logic circuits compared the duration of each set of three and used this to select the third fastest wheel at any moment. This was to help it ignore the worst of any slippage under power. For example, if a wheel temporarily lifted off the surface, it would spin up because power was being supplied to its motor in the absence of traction. This meant it would generate excessive odometer pulses that were not representative of the Rover's true speed. Likewise, if a wheel were to drag for some reason, perhaps a faulty bearing, then it would generate too few pulses. Hence the third fastest wheel was selected as likely to give the most reliable reading.

In order to understand how wheel rotation converted to distance travelled, engineers carried out tests and determined that under a normal load, and taking compression of the tyre and wheel slippage into account, one rotation of the third fastest wheel would equate to about 2.2m.

Engineers named the sets of three pulses coming from the third fastest wheel as 'distance increments' and they would be used throughout the subsequent processing. Since there were nine pulses per rotation and hence three 'distance increments', one increment therefore represented 0.735m (i.e. one third of a wheel rotation). The numeric display was marked in tenths of a kilometre and required one electronic signal to click up 0.1km (100m). This could be generated by counting 136 distance increments, which is what the SPU did. Actually, 136 distance increments is slightly less than a tenth of a kilometre, but this error was well within the expected accuracy of the system considering that the Rover would be driving over rough and varying terrain.

A much more complex task for the SPU was to process the information from the gyroscope unit and, along with the distance increments, determine the Rover's current position on the ground with respect to the starting point. With this, it could calculate the range and bearing back to the LM.

The three wires that came from the gyro, which carried a representation of direction, were electronically converted to two signals that did the same thing, except that the voltages represented the components of direction as north/south or east/west values. These voltages were then turned into numbers, or digitised, so that they could be used in the calculations required in the next stage.

With a digital idea of what direction the Rover was travelling, the SPU could make sense of each distance increment as a distance travelled north/south and east/west. For example, when driving exactly northwest, through the trigonometry seen in the diagram below, a distance increment (0.735m) became a north/south distance of +0.52m and an east/west distance of –0.52m.

While the Rover was in motion, the SPU added up (or accumulated) the north/south and east/west values from each distance increment. As a result throughout the journey it had two numbers that represented the vehicle's current position with respect to its starting point, usually the LM. Those numbers indicated how far north/south and how far east/west of the LM they had

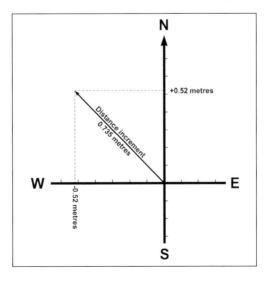

LEFT Diagram of how a northwest 'distance increment' can be resolved into two components of northerly and westerly travel. *(David Woods)*

RIGHT A Rover that travels 1km east and 1km south from the Lunar Module is on a bearing of 135° from its starting point at a distance of 1.4km. To return to the LM, it can follow a heading of 315° for 1.4km. *(David Woods)*

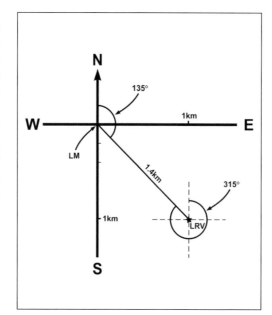

travelled. These were 'Cartesian' coordinates, similar to the x and y graphs used in mathematics, but they didn't suit the astronauts trying to get around the surface in a hurry.

When pilots fly around the skies, they generally think in terms of bearing and range to their next waypoint rather than reading their coordinates at any one moment. Since the navigation system was to be operated by pilots, it would need to tell them their position in their own language. Bearing and range are a form of polar coordinates, so the SPU's next task was to convert the Cartesian coordinates into this description of the Rover's position. For example, if they travelled 1km east of the LM and 1km south, they would reach a point that could also be expressed as a bearing of 135° from north at a distance of 1.4km (as shown in the diagram).

Finally, the SPU had to transform the polar coordinates from ones centred on the LM to ones centred on the Rover. So in the example above, in order to return to the LM they would have to drive on a bearing of 315° for a distance of 1.4km.

The system worked on paper, but would it work in practice on a Lunar Rover traversing rough terrain? Before manufacturing it into the prototype-rovers, Boeing decided to conduct some pre-prototype tests on an Earth car. Their experimental kit was installed into an SUV-type vehicle of the time called a Travelall, and magnets were affixed to its wheels in order to operate switches to simulate the odometer pulses. The driving course selected was in the Merrium Crater area of Flagstaff, Arizona.

To make the test more realistic, the Travelall vehicle had its windows blacked out, forcing the driver to guide solely using the LRV nav system. To avoid crashing into things, a TV camera was mounted on the hood to provide a limited view of the route ahead.

Pleasingly for the engineers, after seven drives around a surveyed course, the results showed that in real-world situations, the navigation system operated better than hoped. But it was essential to accurately set it up at the start of each drive; just as the astronauts would have to do on the Moon.

> *The 1G Rover navigation system was calibrated for use with wire wheels; however, the navigation errors incurred when pneumatic tyres were used were small.*

Nav initialisation

For the navigation system to make any sense during a drive over the lunar surface, it had to be properly initialised at the start of a day's outing. Just like on a modern car, there was a button on the Control and Display panel that allowed the trip counter to be zeroed. This System Reset button also zeroed the range and bearing-to-the-LM displays, and cleared the internal accumulators in the SPU.

Next, the directional gyro had to be aligned with the Moon's geographical north, the North Pole around which the Moon rotated. Without a global magnetic field, the only reference available to determine the direction of lunar north was the Sun. For this, there was a fold-up 'Sun shadow device' at the top of the Control and Display panel, adjacent to the compass rose. When deployed, this allowed the shadow of a needle to fall on a small scale that measured degrees.

To find lunar north, the astronaut had to park the LRV with the Sun at his back, which he could judge by observing his own shadow. He then deployed the shadow device to see

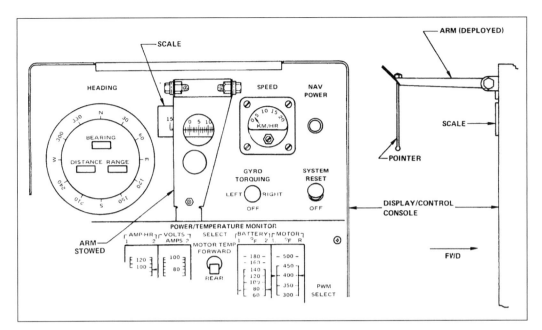

LEFT Diagram showing the stowed and deployed positions of the Sun shadow device. *(NASA)*

BELOW LRV-2's Sun shadow device is shown in its deployed position. *(NASA)*

where its shadow landed on the scale. This gave an angle that indicated which way the LRV was pointed with respect to the Sun. One complication was the slope on which the vehicle was likely to be parked, which could affect the result. For this, a simple pendulum instrument was provided to measure the Rover's pitch and roll angles.

Once all three angles – Sun, pitch and roll – had been radioed to Mission Control, they were combined with a knowledge of where the Sun actually was in relation to the Moon to yield the LRV's true heading. Mission Control passed this true heading figure to the astronaut who could then use the gyro torquing switch to rotate the gyroscope mechanism until the compass rose displayed their current heading. With the gyro now pointing at the Moon's true north, the drive could begin. At each geology stop, when the astronauts set about their science goals, the navigation system remained powered in order to ensure the gyro, counters and accumulator retained their state throughout the EVA. At the end of the day, the system was powered down. Then after a night's rest, the crew would switch it back on and reset it at the start of the next day's travels.

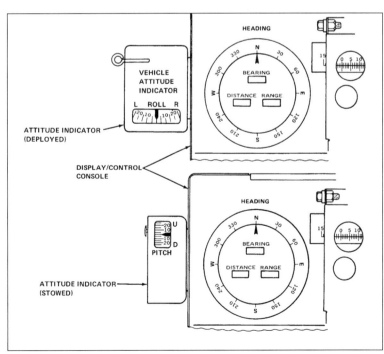

RIGHT Diagram showing the stowed and deployed positions of the attitude indicator. *(NASA)*

LEFT A composite of four images from Apollo 17 with Gene Cernan standing by LRV-3. Attached to the front of the chassis is the Lunar Communications Relay Unit (LCRU) with its gold-coloured blankets hanging forward. The TV camera is mounted to the left and the high gain antenna is on the right. Visible behind the TV camera is the low gain antenna and both it and the high gain are aimed towards Earth. *(Harrison Schmitt/NASA/David Woods)*

Communications

By driving far from the LM, and very likely out of sight of the lander, the astronauts could not depend on the Lunar Module to provide a communications link to Earth. (Refer to the Haynes *Apollo 11 Owners' Workshop Manual* for the full description of direct LM–Earth communications). Instead, the Rover had a complete self-contained communications package capable of direct live TV broadcasting to Earth. This was the Lunar Communications Relay Unit (LCRU), which was carried to the Moon inside an equipment bay of the Lunar Module's descent stage and then installed on the Rover by mounting it on two pins that made it stick out at the front of the forward chassis.

The LCRU sat at the centre of a network of radio links. The astronauts on the surface spoke to each other using VHF radio (250–300MHz). The LCRU picked up these transmissions and relayed them to Mission Control. It also received transmissions from Earth and relayed them to the astronauts. Depending on the circumstances, this two-way link to Earth at a frequency of just over 2GHz (in the 'S' radio band) used either a small, helical-feed low gain antenna or a larger umbrella-type high gain dish antenna.

The high gain antenna was a very distinctive add-on to the Rover. It was taken to the Moon while furled like an umbrella, and then opened up into its final shape once it had been installed

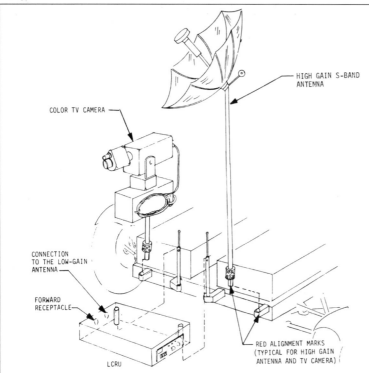

LEFT Diagram to show the placement of the communications equipment on the front of the Rover. *(NASA)*

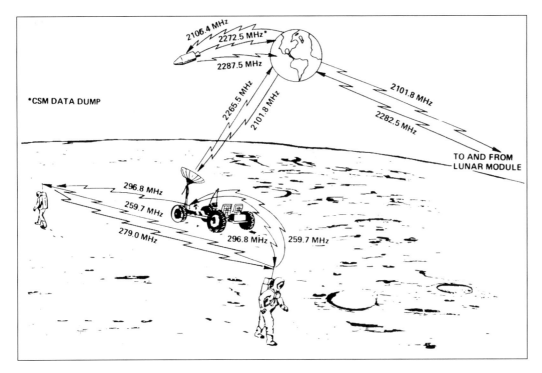

LEFT Diagram to indicate the communications links of an Apollo mission during a Rover drive. *(NASA)*

BELOW Jim Irwin works at LRV-1 at the end of Apollo 15's first drive. The dish shape of the unfurled high gain antenna is apparent. *(David Scott/NASA)*

BELOW LEFT Diagram of the high gain antenna in its folded state, with view (inset top left) through sighting scope, including graticule markings. *(NASA)*

on the left side of the forward chassis (see Chapter 6). It was needed because television was notoriously greedy of bandwidth and required a strong signal to the large receiving dishes on Earth, otherwise they would receive an excessively noisy picture. By using a dish-shaped reflector to shape the antenna's beam to be as narrow as possible, the signal was concentrated like a torch beam. Because such a narrow beam had to be aimed within 2.5° of the centre of the Earth's disc, which spans 2° in the lunar sky, the assembly included a sighting scope through which the space-suited astronaut peered to centre Earth in a graticule.

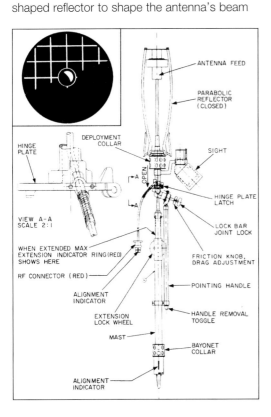

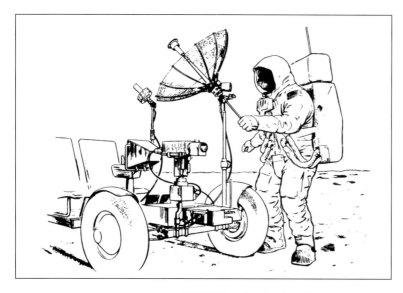

ABOVE Drawing to show how an astronaut aimed the high gain antenna. *(NASA)*

ABOVE Apollo 16 landed near the centre of the Moon's near side and this meant that Earth was just about directly above their heads. Though John Young had to adjust the aim of the high gain antenna, as he is doing in this photograph, the low gain antenna's aim never needed readjustment for this mission. It was always near enough. *(Charlie Duke/NASA)*

LEFT An LCRU is attached to the front of the 1G Rover, wrapped in Mylar thermal blankets. Its carry handle is visible sticking out of the very front. Charlie Duke and John Young pose for the press on 22 December 1971 during training. *(NASA)*

Maintaining an accurate aim at Earth was impossible when the Rover was moving, so before they left a site the crew always turned off the TV camera and switched audio communication to use the low gain antenna mounted on the commander's side of the Control and Display panel. This antenna's beam was much wider at about 30° and so was much more tolerant of pointing errors. Its mount included a handle and mechanism to allow it to be rotated to approximately point at Earth and then be locked in position.

Among the further abilities of the LCRU was that it could pass signals from Earth received through the high gain antenna to allow the TV system to be controlled from Houston. There were also arrangements for signals from biomedical sensors on the crew and engineering sensors from the LCRU itself to be sent to Earth for monitoring.

The LCRU's power requirements were quite high. Across the length of a lunar stay, it required almost as much energy as the LRV itself. It consumed 124W with the TV operating and 74W without television.

Although the LCRU could be powered from the Rover's batteries, engineers decided to give it its own interchangeable battery. As noted in

RIGHT **If the Rover were to fail and the astronauts had to walk back to the LM, they could keep in contact with Houston by carrying the LCRU and making sure that as they walked, the low gain antenna was kept roughly aimed at Earth.** *(NASA)*

Chapter 3 this made it capable of operating entirely independently so that if the astronauts found themselves having to walk back to the LM, they could detach the unit and take it with them in order to maintain communications with Earth. Refitting was the reverse of removal. For this, there was a carry handle and a receptacle for the low gain antenna. For LRV-1 on Apollo 15, three LCRU batteries were supplied: one for each day.

Television

The Apollo J-missions fully integrated television technology into the philosophy of the LRV. By freeing the camera from a static mount near the LM, flight controllers and scientists in Houston, and indeed the whole world, could join the astronauts on their exploration of the Moon. However it was not always so.

On Apollo 11, NASA engineers and managers

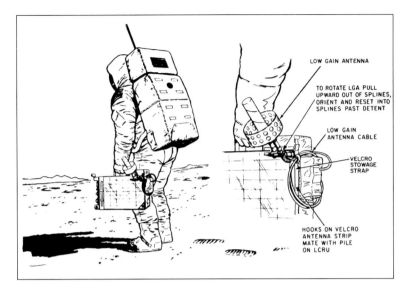

BELOW **An exploded view of the LCRU seen from below shows its major components, including three wax packages for thermal control.** *(NASA)*

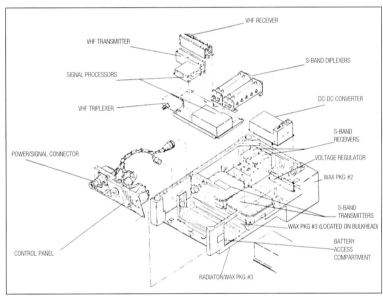

BELOW **A frame from the Apollo 11 black and white TV camera. Although this footage has been restored nearly 40 years after it was shot using the best master footage available, it is still of poor quality.** *(NASA)*

RIGHT **Westinghouse engineer Stan Lebar with the television cameras that were used on Apollo 11. The unit on the right was the black and white camera for the lunar surface. The colour unit on the left was only used in the spacecraft and was never taken to the surface.** *(NASA)*

105

NAVIGATION AND COMMUNICATION

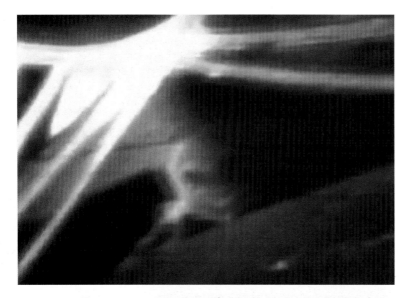

could barely justify TV in terms of its scientific or operational value. They only grudgingly acceded to a simple black and white camera whose scan rates were non-standard. Deficiencies in the camera and in the conversion process to broadcast standards, meant that viewers around the world saw a poor, contrasty image with strong ghosting artefacts. However, the PR success of having live pictures from the Moon made NASA realise the power of television in bringing its achievements to a wider public.

Everyone looked forward to the pictures that would come from the colour camera that was taken to the surface on Apollo 12 and it worked well until it was inadvertently pointed at the Sun. The strong light and heat burned the camera's photosensitive tube, which meant that the rest of the moonwalk had to be carried out with voice alone. For those watching the missions from Earth, things improved little when Apollo 14's colour camera displayed excessive image blooming. Worse, it was unable to follow the astronauts when they embarked on a potentially spectacular 1km trek to a large crater, the scientific centrepiece of the mission.

By this time, science was firmly in the driving seat and the geology team realised that if they were to fully support their crew, they would need to see what the astronauts were seeing. But NASA could not afford to have an astronaut leave his tasks to look after a live camera.

ABOVE This frame from Apollo 12's colour TV camera was taken moments before the unit was rendered useless by being inadvertently aimed at the Sun. *(NASA)*

RIGHT Apollo 12's hapless colour TV camera on its tripod on the Moon. *(Pete Conrad/NASA)*

BELOW Astronaut Alan Shepard takes a golf shot in this TV camera frame from Apollo 14. The bloomed image of the sunlight reflecting off his 'club', a tool handle, is an example of the camera's deficiencies. *(NASA)*

BELOW On the right of this Apollo 14 TV image, three indistinct images can be seen heading off for a kilometre-long trek: Alan Shepard, Ed Mitchell and their hand-drawn cart. The camera remained attached to the LM and could only transmit this one hazy view with the sunlight flaring in the lens. *(NASA)*

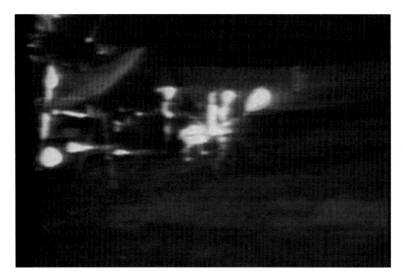

ABOVE **Director of the Marshall Space Flight Center, Wernher von Braun with an example of the RCA colour TV camera of the type that was used on the Rovers.** *(NASA)*

It was a young engineer in the Engineering and Development Directorate, Bill Perry, who came up with the idea of a remotely controlled camera on the Rover. In order to perfect its design, Sam Russell created an impressive mockup of the lunar surface and strove hard to get the strong unidirectional lighting right.

All the lessons from the previous missions were applied to the Rover's TV system to produce a unique, compact, mobile, remotely controlled television outside broadcast unit that could transmit live TV from the surface of the Moon wherever the Rover happened to stop. Comprising the TV camera together with its control apparatus, this was known as the Ground-Commanded Television Assembly (GCTA) and it was built by the RCA Corporation.

Like its contemporaries, the camera used vacuum tube technology which came before solid-state CCD or CMOS sensors as used on all modern electronic cameras. An image with a 12.75mm diagonal was read from a photoconductive surface using an electron beam in a silicon intensifier target (SIT) tube, a type which had been developed for low-light applications in the military while also being resistant to image burn. Conventional electronics then used this to produce a standard US black and white television signal of 60 pictures per second that was sent to Earth via the LCRU's S-band link.

Colour cameras work by scanning a scene in the three primary colours; red, blue and green. Whereas other colour cameras used at least three imaging tubes to analyse the scene simultaneously, the LRV camera's single tube got around this by being placed behind a spinning filter wheel which allowed it to scan the colours one after the other. The filter wheel had six filters in the sequence; red, then blue, then green, followed by a repeat of the sequence. By synchronising the wheel's rotation with the tube's scans, the sequence of TV images coming from the camera became a sequential analysis of the scene in red, then blue, then green. As the picture rate was 60 per second, all three colours would be scanned 20 times per second and because the wheel had two filters

ABOVE **The TV camera of LRV-3 with the box-shaped Television Control Unit underneath and a sunshade fitted to the front of the lens. This was the Rover's last outing. Harrison Schmitt is taking still photos of the landscape with a film camera and telephoto lens.** *(Gene Cernan/NASA)*

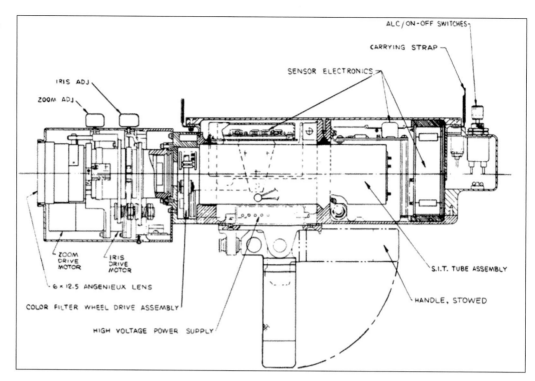

RIGHT A diagram of the internal systems of the Rover's TV camera. *(NASA)*

BELOW The TV camera on Apollo 16 with sunshade. *(Charlie Duke/NASA)*

of each colour it was spun at ten revolutions per second.

Upon reception on Earth, the signal was processed to become a standard NTSC colour signal for widespread TV broadcast distribution. First its timing was restored. Analogue TV signals were built up line-by-line for each picture. The timing of those lines had to be extremely accurate and properly synchronised with the rest of a TV studio to be of any use. If the signal was coming from another world that was moving with respect to ours, and from a camera whose timing was not synchronised with anything, the pictures would be useless. To fix this, they were recorded onto a reel-to-reel

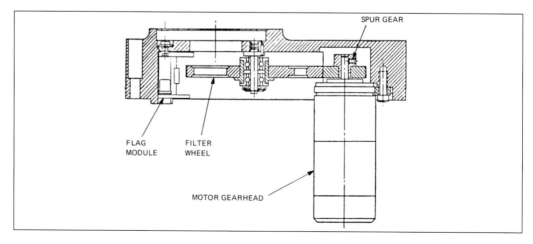

RIGHT Cross section diagram of the colour wheel mechanism. *(NASA)*

RIGHT A colour wheel with two sets of three dichroic filters. *(NASA)*

videotape recorder that synchronised itself to the video from the Moon. The tape was then fed immediately to a replay machine that was synchronised to TV standards. Finally, magnetic discs were used to store the sequential colour images and play them back concurrently so that they could be combined into a composite colour TV signal.

The TV camera had a 6-times zoom lens by Angenieux of France whose focal length range was 12.5–75mm. Its aperture range was f/2.2 when fully open, down to f/22 when completely stopped down. Both zoom and aperture could be adjusted either by the astronaut by way of controls sized for a suited hand, or by remote control from Earth via stepping motors. Focus control was coupled to the zoom and most subjects beyond a distance of 1.7m could be brought into focus merely by zooming.

When the first of these RCA cameras went to the Moon on Apollo 15, it was quickly realised that the combination of even the tiniest amount of dust on the lens and very strong sunshine made many of the shots look washed out due to sunlight being scattered into the lens. The designers had not realised the importance of a sunshade but added one to both of the cameras that flew on subsequent missions. When the astronauts took 16mm cine footage from on board the Rover during drives across the Moon, the TV camera was in the field of view and the sunshade was clearly visible. It was a significant reason for improvements in picture quality across the final Rover missions.

Another factor was NASA's use of a new proprietary video enhancement system from a company called Image Transform. NASA sent the live television from Houston to Image Transform in California where it was processed in real time and returned.

Below the camera sat the Television Control Unit, a box with an altitude/azimuth mount to

LEFT A TV frame from Apollo 15 coverage. The sunlit flank of the 4,500m Mount Hadley seen through the haze caused by dust on the front surface of the lens. *(NASA)*

RIGHT This still frame from Apollo 16's 16mm cine film coverage shows the TV camera in the foreground with its sunshade. *(NASA)*

ABOVE A frame from LRV-3's TV camera shows Gene Cernan adjusting the aim of the low gain antenna. *(NASA)*

RIGHT Apollo 17's TV camera with its pan and tilt mount and Television Control Unit. *(Harrison Schmitt/NASA)*

pan and tilt the camera remotely. It contained the electronics to decode commands from Earth, distribute them to the correct system (lens, camera or motorised mount) and it also routed the video signal from the camera to the LCRU for transmission to Earth. The whole assembly was mounted on a pole that installed into a fitting on the right-hand side of the Rover's forward chassis.

For the flight from Earth, the TV camera was mounted on a deployable panel in the side of the LM so that the crew's initial descent down the ladder and their first steps could be televised. The commander pulled a handle near the top of the ladder to open the panel. At this point, the camera was connected to the LM's electronics by a 30m cable. Once the astronauts were out and on the surface, the camera could be moved onto a tripod to view the deployment of the Rover. Only then was it disconnected from the LM, installed on the LRV's chassis and connected to the LCRU. Once it was mounted on the LRV with a good S-band connection with Earth established, Mission Control could take control of the camera's pan, tilt, aperture, zoom, focus and automatic exposure mode.

This was the responsibility of INCO flight controller Ed Fendell, who took great pride in his operation of the camera while the astronauts got on with their work. He remembers how

LEFT Flight controller Ed Fendell at the INCO console from where he operated the Rover TV camera. *(NASA)*

timing was everything. "I'd sit in Mission Control and use a series of buttons which sent signals to move the camera on the surface." However, the distance between the Earth and the Moon meant there was a delay of around 3sec between him pushing a button and sending the signal, to him seeing the camera move back on Earth. "So I had this rather annoying delay; it took some time to get used to. The big problem was we never knew which way the astronauts would start walking – it must have looked like a comedy of errors at times!"

The idea of a camera on the Lunar Rover paid off. Flight controllers could watch the astronauts work on the surface, keep a log of their activities, and help them when problems arose. Geologists could scan the surrounding terrain and help the crew come to understand the story of the place they were exploring. At times, they could even suggest rocks for an astronaut to pick up. Like no other feat of exploration before, millions of TV viewers could become part of the event and see the same unearthly views the astronauts were seeing, thanks to the Rover and its camera.

ABOVE Tony England at the Capcom console in the Mission Operations Control Room during the Apollo 16 mission. Standing behind him is Deke Slayton, grounded Mercury astronaut and the boss of the other astronauts. Pictures from the Rover's TV camera are displayed live at the far end of the room to allow everyone in the room to follow the mission's progress. *(NASA)*

BELOW A view across the Mission Operations Control Room shows David Scott and Jim Irwin working on the Moon. In the foreground are Capcom Joe Allen and backup commander Dick Gordon. *(NASA)*

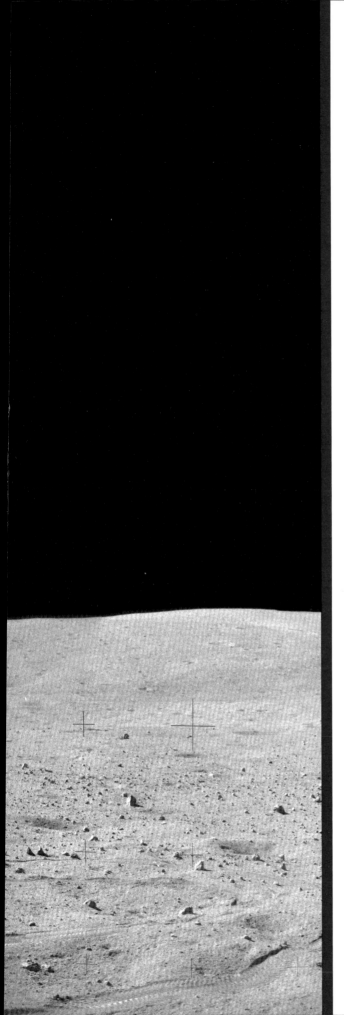

Chapter Five

Crew station & Control and Display Console

It is 23 April 1972. John Young and Charlie Duke are gingerly negotiating an extensive impact debris field emanating from South Ray Crater on the Cayley Plain in the lunar highlands some 270km south of the Moon's equator. Even at the relatively slow speeds at which John is nudging the Rover forward, the two of them are being heaved from side to side as they veer around the football-sized rocks which they estimate litter 40–50 per cent of the ground.

OPPOSITE LEFT Panorama constructed from Charlie Duke's Station 9 images, showing John Young with the dust brush at the front of the Rover, probably just before he moves around to dust the TV lens. *(Charlie Duke/NASA/David Woods).*

RIGHT One-sixth-g parabolic flights with a mockup of the Lunar Rover crew station, used to practise techniques for Rover ingress and egress. *(MSFC/NASA)*

Both men feel vulnerable at times, conscious of their top-heavy backpacks dragging them out of the vehicle as it pitches and rolls. But, despite the fact no one has come this way before, they are confident that the Rover and its all-important 'crew station', which is currently protecting them, won't fail.

As accomplished military test pilots, both are used to piloting prototype vehicles into unknown territory in the pursuit of engineering perfection. It's a process that they've been engrossed in with the Lunar Roving Vehicle for the past two years, to ensure that the machine they now sit on will ferry them safely across exactly this sort of testing, uncharted terrain.

Before leaving for the Moon they'd both made one last flight in NASA's KC-135 flying one-sixth-g parabolas to make sure their seat belts will function perfectly on the voyage ahead. Now, 400,000km from Earth, the LRV engineers' attention to detail is paying off.

Getting it right first time

Though many people may not think of it as such, the Lunar Roving Vehicle was a spacecraft, and the first one in history to be manned on its maiden mission. There was no dress rehearsal flight to learn from, and no opportunity to make improvements before the first human beings rode it on the Moon. So to fine-tune the vehicle's ergonomics Marshall and its contractors conducted regular crew station reviews with the astronauts themselves.

When the prime crew wasn't available to participate in the design process their backups would step in, and when they were too busy the engineers themselves donned space suits. Many of these human factor reviews took place on board NASA's infamous KC-135 'vomit comet' to simulate the one-sixth-g environment of the Moon.

In mid-1971, with both the prime and backup crews on other business, it was manned systems integration engineer Mike Vacarro's turn to assess climbing on and off the Rover under simulated lunar gravity. "We'd only get 20 seconds or so of one-sixth-g on each dive," he explains. "We flew 50 parabolas that day! We were up there quite a while. In fact I have more zero-g time than those astronauts! I had even more than the Original [Mercury] Seven," he boasts.

There was a downside to these thrilling rides, admits Vacarro. "Many of the astronauts would get sick. It wasn't so bad if they weren't wearing the space suit, but several of them got sick inside the space suit and that was a mess, to clean on out."

Enduring the discomfort of these nausea-inducing flights was vital to understanding an astronaut's limitations in his pressure suit. And this work would help to refine the parts of the vehicle that would enable a human being to ride it in comfort across the Moon.

Crew station

The Rover's so-called 'crew station' comprised everything that the vehicle required to accommodate an astronaut. It included controls to drive and monitor the vehicle, hand and toeholds to help him climb on board, floor panels to stand on, seats to sit on, seat belts to restrain him, and foot and arm rests for comfort. Finally, like every useful car, there was also a place to stow his stuff: tools, experiment kit, and the samples that he collected. Designers also added lightweight fiberglass fenders to protect the astronauts from lunar dust kicked up by the wheels (see Chapter 2).

Hand controller

The Lunar Rover was a true drive-by-wire vehicle because almost all of the drive instructions from the astronaut were communicated to the drive mechanisms of

RIGHT AND CENTRE Layout of the LRV's crew station. *(MSFC/NASA)*

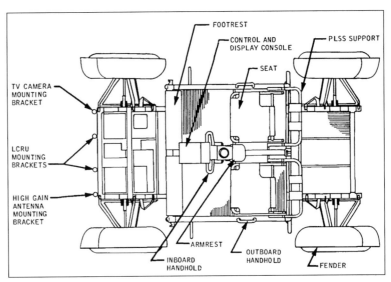

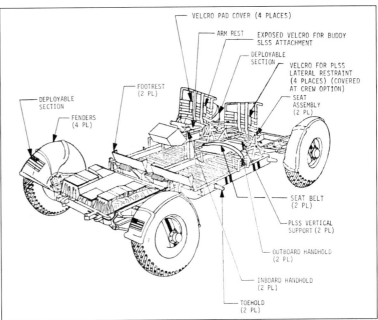

the vehicle through electrical signals. Only the brakes had a direct mechanical connection to the control handle via cables. It wasn't the first vehicle to incorporate this sort of technology; another wheeled Apollo vehicle called the Crawler, used to carry the rockets to the launch pad, was also a drive-by-wire vehicle. But the LRV was the first personal car-sized vehicle to carry this technology, so common in the acceleration systems of similar sized cars today.

NASA had stipulated that the Rover should be controllable from either seat, implying the need for two steering wheels. But this would seriously obstruct the ingress and egress of two crewmen wearing bulky pressure suits. In seeking a different solution, Boeing studied how the astronauts interacted with controls whilst wearing space suits. With such restricted mobility, foot pedals were quickly ruled out. This led naturally to using some sort of universal hand controller to generate steering, speed and braking commands to the Rover's systems. By locating a single device between the astronauts, either of the two men could operate the vehicle with only one hand. And a centrally mounted controller would ensure clear access to their seats.

The hand controller would have to be a marvel of multifunction, and Boeing initially conceived it as a pistol grip design. "It had a carved handle just like a control stick in a modern airplane," says Gene Cowart, Boeing's chief engineer for the Rover. "Either astronaut could run this Rover. All he had to do was push forward and he went faster. Pulled back it set the brakes. If he pushed it to the left it steered left, if he pushed it to the right it steered right. If he wanted to back up there was a switch right on the side that he could flick, reverse the polarity and the thing would back up. It was really quite straightforward and probably more efficient than a [steering] wheel."

RIGHT MSFC programme manager of the LRV, Sonny Morea, checks an early mock-up of the crew station, testing a pistol-grip hand controller. This was later changed to the T-handle. *(NASA/MSFC)*

DRIVING THE ROVER

The following instructions for operating the Apollo Lunar Roving Vehicle are presented here in a form drawn very closely from NASA's original LRV owners' manual.

Speed control

Tilting the T-handle away from you on its palm pivot allows control of the LRV's forward speed. A constant muscle force of about 6 inch-pounds (or 0.7 newton-metres) is required to tilt the hand controller beyond its neutral position or 'dead band'. Tilting further from the neutral position about this pivot axis proportionately increases forward speed. A 9° forward tilt corresponds to half the maximum power from the drive motors. Maximum power is achieved by tilting the hand controller further forward to its hard stop position at approximately 14° from neutral.

To decelerate, the hand controller is tilted back towards neutral. To place the vehicle in neutral, the hand controller is returned to within ½° of the 0° rotation position.

To operate the vehicle in reverse, a so-called 'reverse inhibit switch' is placed in the up position and the hand controller tilted backwards about the palm pivot. The degrees of displacement for reverse are identical to forward speed operation.

With the reverse inhibit switch in the down position, the hand controller can only be pivoted forward, thereby preventing the vehicle being inadvertently reversed.

As a further safety measure, to protect the drive motors being switched too fast from forward motion to reverse, logic circuits ensure that power to the motors is not reversed until the LRV is moving slower than 1kmh. Despite this precautionary feature, a quick shift of motor direction with the T-handle can leave the wheel speed sensor logic not fully stabilised; resulting in some wheels being in forward state and some in reverse state upon reapplication of power.

This is because each wheel's speed sensor uses a 'monostable' circuit to generate the 'no-go' signal. A monostable circuit has a tendency to remain in one state – in this case a 'go' state – whilst not receiving any external signals. But whilst pulses are arriving from the wheels, the circuit is forced into its alternative 'no-go' state, preventing wheel motor reverse direction from being selected. But at slow speeds the pulses arrive at such a slow rate that the circuit's 'go' state can occur before the next pulse arrives, briefly permitting a wheel motor reversal for the wheels which have slowed enough. Therefore, while the car is still going forward, albeit slowly, it is possible that only some wheels might have gone into reverse mode.

To avoid this, the vehicle should be brought to a complete stop to ensure that no pulses are being received by the monostable, thereby allowing all wheels to enter reverse mode together.

BELOW AND OPPOSITE The T-bar hand controller created by Mike Vacarro's team as a solution to the hand fatigue problem of the pistol grip. (MSFC/NASA)

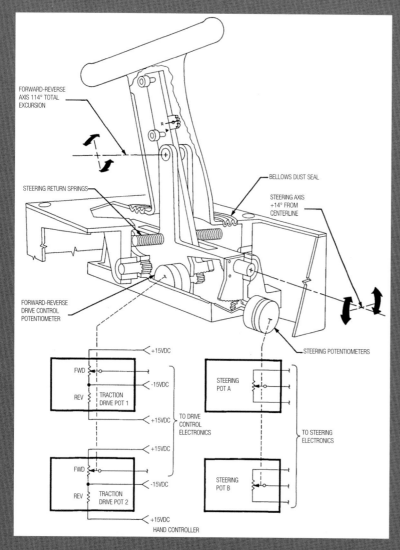

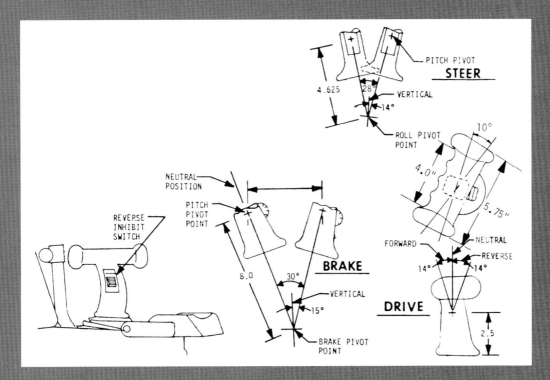

Braking

Pulling the whole T-handle (a bit like a gear shifter in an automatic car) engages its lower pivot, operating the brake cables through linkages to the brake drums on all four wheels. The braking force is applied proportionally as the handle is moved backwards through its full 30° of travel.

This physical connection to the brakes is useful to the astronaut for those occasions when they are required to apply power whilst also braking; such as when pulling away on a hill. In this situation the T-handle could be upper (palm)-pivoted by tilting the T-bar and lower- (brake) pivoted by pulling back on the column simultaneously. But for safety, forward and reverse power is cut when the brake is applied by a lower pivot position of 15° or more.

The parking brake is engaged by pulling the handle back further than 15°. This takes up the slack in the cables and operates the mechanical brakes. A detent lock holds the joystick to act as a parking brake.

To disengage the parking brake, the T-handle is placed in the steer left position. An extra contingency brake release method is provided for use in the event of the brake failing to release when moved to the steer left position. This is achieved by moving the T-handle to its full rearward displacement and then pulling a release ring (found below the wrist rest section of the armrest). Once the brake disengages, the release ring can be released too.

For details of how the brakes work see Chapter 2.

Steering control

Pivoting the hand controller left or right about the roll pivot point proportionally changes the wheel's steering angle. Applying a muscle force of 1 inch-pound (0.12 newton-metres) to roll the hand controller beyond the neutral position will begin to change the steering angle. The effort required to increase the displacement angle about the roll pivot point rises linearly until, at a displacement of approximately 9°, a soft stop is encountered. An increase of 5 inch-pounds (0.6 newton-metres) is required to pivot the hand controller beyond this point to its hard stop limit at the full 14° displacement.

The hand controller is spring loaded to return to its neutral steering position when released, automatically realigning the wheels with the LRV's centreline. Like the throttle control, the steering control also has a ½° neutral dead band on either side of the 0° position.

For details of how these movements of the T-handle actually control the double Ackermann front and rear steering, see Chapter 2.

ABOVE A slightly elevated view of LRV-1 in early 1971, prior to delivery to NASA, giving a good view of the seats and crew station layout. *(MSFC/NASA)*

Simple as it sounded to drive the Rover with the control handle, building such a universal control stick posed difficulties. "The conductors which came up through the column from down below had between 130 and 140 electrical cables," points out Vacarro. "They had to connect to all of the systems on the Rover through this item." Through the first half of 1970 Boeing worked tirelessly to integrate it into the vehicle.

But during this time it became apparent that the pistol grip was difficult to use in a pressure suit. "They had designed it so that you moved it like a control stick in yaw with your wrist," explained Apollo 16's commander John Young to aerospace historian Anthony Young, for his 2007 book *Lunar and Planetary Rovers.* "The wrist would be so tired you wouldn't be able to drive after two or three minutes."

Just months before the final Qualification Test Unit was to be delivered to NASA at the end of 1970, Vacarro was asked to devise an alternative control handle. "It was a 'crash' programme, in which we were told we had a certain number of hours to get that design completed and tested to the satisfaction of the astronauts that were going to fly the remaining lunar missions," he remembers. "There was a lot of pressure on us to get it done."

Vacarro tried out various hand controllers, such as a bayonet-type grip and a knob, but these still needed to be gripped continuously, causing fatigue after just a few minutes. Further modifications resulted in a T-shaped handle which required no gripping to operate it. "There was no reason to have to stop and relax your hand," points out Vacarro. "With the thumb on one side of the T-handle they could always relax the four fingers on top, but still apply pressure to be able to control the Lunar Rover."

Racing to complete the design within the time constraints, the new T-handle was thoroughly tested using the elaborate Rover simulators at Marshall to have it ready by November 1970, just in time for integration into the Qualification Test Unit required to manufacture LRV-1 for its flight to the Moon the following summer.

For astronaut comfort, a fibreglass armrest was also incorporated into the design to support the crewman's forearm whilst operating the controller.

The T-handle device had three pivots: an upper so-called palm pivot, a lower brake pivot and a roll pivot. Tilt the T-handle forward to engage the palm pivot and the Rover went faster. Tilt it back and the vehicle slowed down. Pull the whole T-handle backwards to engage the lower pivot and it applied the brakes. Lean it left or right on its roll pivot and the vehicle steered in the indicated direction (see box on previous spread).

Seats

The Rover's seats were mounted side by side, with the commander sitting on the left and using his right hand to control the Rover. They were made of tubular aluminium frames, spanned by a nylon webbing that gave the appearance of a lightweight beach chair. They folded flat onto the centre chassis when the vehicle was stowed, and were pulled upright after the vehicle had been deployed onto the lunar surface (see Chapter 6).

The side-by-side seating arrangement allowed both astronauts to see the front wheels and the central joystick controller; allowing the Rover to be driven from either seat. However, the standard operating procedure was for the commander to grip the handle to steer, accelerate and brake (see Chapter 7).

Perhaps uniquely for a wheeled vehicle, the seats had to accommodate the bulky backpacks of the astronauts' Portable Life Support Systems (PLSSs). Along with the space suit, these 'wearable spacecraft', as some engineers referred to the combined suit and PLSS, weren't designed for sitting in. Under internal pressure and weighed down by the backpack, the suits had a tendency to straighten out, forcing the seated astronauts into a reclining position which was far from ideal for driving.

To help hold the crew in a more upright ride position, and also to reduce any significant lateral motion of the heavy life-support backpacks, Velcro patches were positioned on the seat backs to attach to corresponding patches on the PLSS.

To facilitate access to the toothed-wheel temperature control dial on the lower right side of the backpack whilst an astronaut was on the Rover, each seat had a hole out of the seat cushion to avoid the nylon webbing becoming entangled in the dial. This minor detail seamlessly integrated the astronaut's existing life-support systems into the vehicle, whilst also reducing the vehicle's all-important weight!

During one-sixth-g tests the astronauts and engineers identified ways to reduce the time required to board the vehicle from as much as 10min to less than 3min. Some astronauts even got it down to a matter of seconds, finding that the very best way to take their seat on the Rover was to hop backwards straight into it. During their three days on the Moon many of the astronauts would become quite adept at this hop, landing perfectly in their seat in a single bound (see Chapter 7).

Seat belts

A seat belt was provided for each seat to restrain the astronaut when the vehicle turned a sharp corner or hit a bump. Like the seat webbing, these were constructed of nylon.

At the belt's end was a hook that was secured to an outboard handhold. A buckle adjusted the belt length, and a stretch section of the belt permitted normal fastening and release. Final length adjustment for each individual astronaut was accomplished during a suit pressurisation test prior to the mission; when each crewmember's length was marked by white thread stitching at the point where the belt looped over the buckle.

Simple as this design sounds, the seat belts were difficult to develop, as Mike Vacarro explains. "When they were in the suit with their helmet on they couldn't see their stomach or the buckle, so they couldn't grab it."

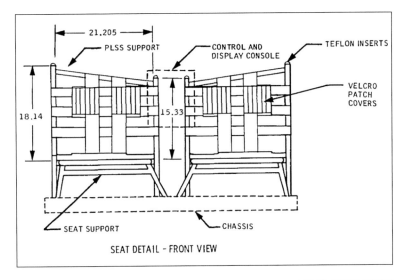

ABOVE Seat front view showing Velcro attachments for PLSS backpacks. *(MSFC/NASA)*

RIGHT Plan view of the seats, showing the webbing removal in the top left corner, to ensure easy operation of the astronaut's PLSS backpack temperature regulation wheels. *(MSFC/NASA)*

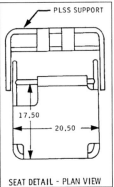

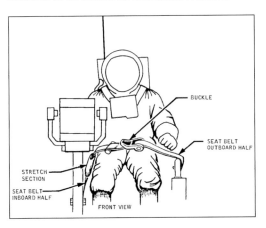

FAR LEFT Seat belt detail. *(MSFC/NASA)*

LEFT Harrison Schmitt seated in LRV-3 during training, illustrating just how impossible it was to look down to their laps with their pressure suits on. *(MSFC/NASA)*

RIGHT David Scott conducts a crew station review, examining the seat belt (probably not at the time of the story noted here). *(MSFC/NASA)*

All this made the process of fastening their seat belt, which was already difficult through a restrictive pressure glove, even more challenging. "We couldn't use a buckle interface like you see on an airplane because you could not physically see to buckle it, and could not feel to buckle it," adds Jim Sisson, Marshall's LRV chief engineer.

Despite trying several configurations, the ground crews just couldn't get the seat belts to work well enough for the first crew who would have to use them on the Moon.

For Apollo 15, Sisson and Vacarro remember that mission commander Dave Scott wanted the seat belt changed. Scott had made a comment about wanting one of the restraints reversed in the other direction. "At that point in time we were very close to starting on [stowing for] the Lunar Module and I didn't want any changes," recalls Sisson. "So I took exception to changing it. At the final crew station review Dave looked at me and said, 'You didn't change the strap.' I said, 'No I didn't change the strap.' Well his crew station chief was George. And he looked at George and said, 'George, give me the scissors.' George slapped the scissors in his hand and Dave cut the restraint. So I didn't have any choice but to change it. So when I changed it I put it back in the direction he wanted. It wasn't funny at the time but it is now."

Scott had been right to force this fix. Sisson remembers him later pointing out that an astronaut's time on the Moon is worth something like $50,000/min, and so it was a big waste of money if it took too long to get the seat belt fastened.

In subsequent ground checks at 1-g before flight, the belts performed to the crew's satisfaction and full suit and pressurisation tests were carried out as planned to set the belt length for each crewman.

But what no one spotted was that once on the Moon, under weaker gravity, there was less suit compression when they sat in the Rover seats and the belts were too tight. "Nobody anticipated the one-sixth-g effects," reflects Scott 40 years on. "We managed, although Jim had more difficulty with his than I did. MCC [Mission Control Center] understood the problems at the time, so when I asked for a stop to fix the seat belts that was ok; but we knew no more time or stops for geology!"

This wasn't the only seat belt issue they had during the expedition, and once back on Earth the crew were quick to engage with the engineers to fix it for Apollo 16.

Mike Vacarro was summoned to a post-flight review in Houston only to be reprimanded by Scott for coming up with a system that had cost them so much time on the Moon. Scott reported that he and Irwin were often unable to find the seat belts when they got back on the Rover, and Irwin hadn't even been able to unfasten his own seatbelt; Scott had had to undo it for him, wasting additional time.

The problem wasn't so much with the buckle, but with the fact that once the astronauts had released the seat belt and it dropped to the side, they'd had trouble finding it again.

Vacarro's solution was to borrow a design from a much earlier form of transport. "To me it was like riding into the horse and buggy days. That whip was always ready and available, it was right there. So a vertical standing rod was attached to the crew station, bringing the belt within view and reach of a seated astronaut who was limited to seeing straight ahead. This allowed him to grasp it very easily and to pull it towards him," Vacarro explains. "They didn't have to look for the seat belt, it was right there anytime they needed it. When they went in and out of the vehicle. When they came back it was still there; self-winding back to its normal stowed position."

Foot rests

Even with the Velcro attachment points on the seat webbing attached to the PLSS backpacks to support an astronaut in a seated position, their suits still had a tendency to force them into a reclined position. Crew station designers therefore came up with a foot rest in front of the seat for each crewmember to push their feet against, to help them to brace themselves in an upright seated position.

For launch, each foot rest was folded down against the centre chassis floor and secured by two Velcro straps. On the lunar surface they were deployed along with the rest of the Rover's folding parts.

Handholds and toeholds

Powered four-wheel steering meant that the Rover could turn completely within 6.1m and be steered lock-to-lock left to right in 5.5sec. According to test drivers on Earth, it was a responsive and lively ride and the one-sixth-g tests that the astronauts undertook on the 'vomit comet' quickly demonstrated the need for toeholds in the floor of the Rover to help them to hold on in the lighter lunar gravity.

There were two toeholds, one on either side of the vehicle. As a weight saving measure, they were created from part of the so-called interface tripods that supported the Rover's chassis when it was stowed inside the Lunar Module. After their removal during deployment (see Chapter 6) the centre leg of the interface tripod was removed and secured in a hole on the outside edge of the chassis to form an outboard toehold (see top right illustrations on pages 48 and 131).

In typical LRV style, these 'recycled' toeholds also doubled up as tools to operate the wheel decoupling mechanism and to release the telescoping tubes and saddle fitting on the forward chassis.

Engineers also wanted to be sure there were enough handholds and grab points around the seats to enable an astronaut to mount the vehicle and take his seat with ease. Following exhaustive one-sixth-g tests, it was decided to add a pair of inboard handholds, made of 1 inch aluminium tubing, between the seats to assist in boarding the vehicle.

These handholds also doubled up as payload attachment receptacles for mounting the 16mm data acquisition camera and the low gain antenna of the LCRU.

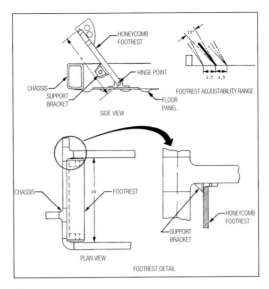

LEFT Foot rest detail. *(MSFC/NASA)*

Floor panels

As noted in Chapter 1, the floor panels in the crew station area were made from beaded aluminium 2219, and were capable of supporting the weight of fully suited, standing astronauts in lunar gravity.

The centre and aft floor panels, along with the handholds, foot rests, and tubular sections of seats, were all anodised with aluminium oxide, providing a heat-reflecting and radiating surface.

Since the underside of the centre chassis floor panels faced outward when the Rover was stowed inside the Lunar Module, it was also covered with special aluminium foil in order to

BELOW Load ratings for the crew station area. *(MSFC/NASA)*

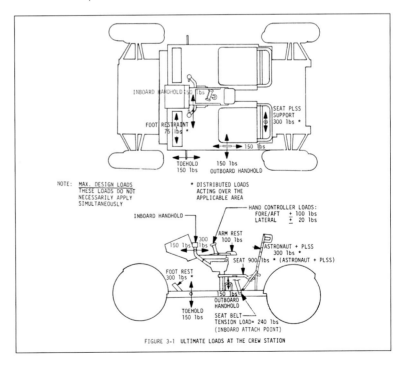

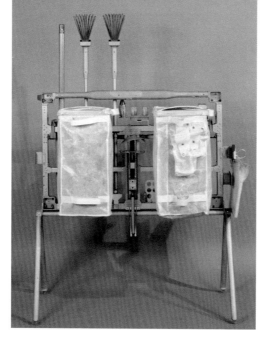

RIGHT **Front view of rear stowage pallet loaded with EVA tools.** *(MSFC/NASA)*

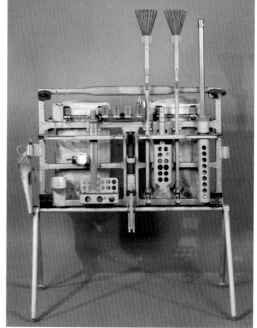

FAR RIGHT **Back view of rear stowage pallet loaded with EVA tools.** *(MSFC/NASA)*

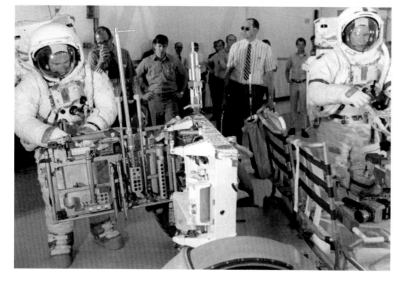

ABOVE **Jim Irwin (left) practises operating the rear stowage pallet whilst Dave Scott works at the seat stowage area.** *(MSFC/NASA)*

BELOW **Plan view of stowage map for the Lunar Rover.** *(MSFC/NASA)*

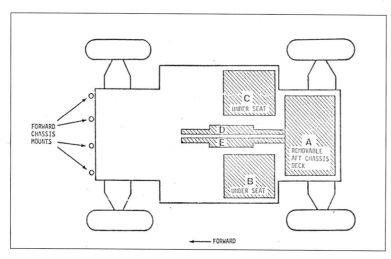

manage the LRV's components' temperature during the flight to the Moon.

Stowage

The 'Aft Pallet Assembly' mounted behind the seats of the Rover had a vertical rack for mounting any experiment kits, carrying tools, lunar samples, stems for the lunar drill, core tubes, and small scientific packages. It was engineered and built by Marshall, and included a hinged gate that helped to secure the tools and permitted additional lunar sample bags to be stowed.

There was further lunar sample stowage beneath the seats. The Hasselblad and 16mm data acquisition cameras were stowed, along with their film magazines, under the seat on the left, and the lunar drill and other experiment hardware under the seat on the right.

Control and Display Console

Along with the Rover's mobility systems, it was General Motors who were responsible for the Control and Display Console which allowed the crew to configure and monitor the performance of the vehicle's power and mobility systems and displayed navigational data.

The external surfaces of the console were coated with heat-resistant paint (Dow-Corning 92-007) for temperature control during the transportation phase, but the face bearing the instruments was painted black to minimise

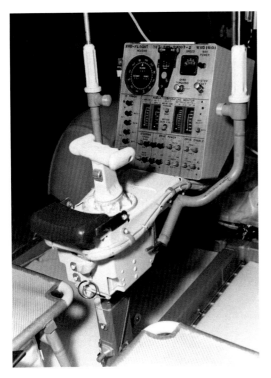

LEFT Display console (probably a non-flight version). *(MSFC/NASA)*

reflection and it was also isolated from the Console by fibreglass mounts in order to limit the transfer of heat. All of the legends were coated with radioactive Promethium in order to provide luminous displays when the panel was in shadow.

Mounted forward of the T-handle and in the astronaut's line of sight, the Control and Display Console was divided into an upper and lower section.

The lower console displayed vehicle monitoring and control systems, and included the master switches for managing the Rover's power (see page 81).

The upper console displayed navigation data (see pages 124–125 and also page 96). It also housed the Rover's caution and warning flags and temperature meters to alert the crew if the batteries or drive motors were overheating, to enable them to take effective action (see page 82).

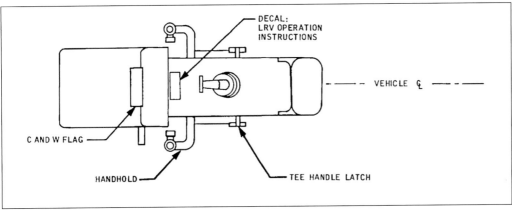

LEFT Plan view of Control and Display Console and T-bar handle. *(MSFC/NASA)*

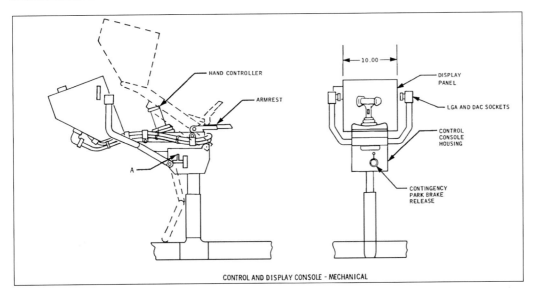

LEFT Side view of Control and Display Console and T-bar handle. *(MSFC/NASA)*

NAVIGATION DISPLAY

The following display console details are presented in the present tense as in NASA's original LRV owners' manual. See Chapter 4 for more details.

Attitude indicator
Mounted on a small box on the left-hand side of the main Control and Display Console is the attitude instrument (see pages 96, 98 and 101). The box displays the vehicle's pitch angle on one side and the roll angle on another. PITCH upslope (U) or downslope (D) is shown within a range of +25° to –25° in 5° increments. ROLL is displayed within a range of 25° left to 25° right in 1° increments. A damper on the side of the indicator can be adjusted to damp out oscillations. The box has a hinge connecting it to the main console. In its folding-in (i.e. stowed) position it reveals PITCH angle and when folded open out to the side of the console ROLL can be read.

Heading indicator
This instrument displays the LRV heading with respect to lunar north. The initial setting and updating of this instrument is accomplished by operating the GYRO TORQUING switch LEFT or RIGHT. *Chapter 4 discusses this procedure in greater detail.*

Bearing indicator
This instrument displays the bearing to the Lunar Module in 1° digits. In the event of power loss to the navigation system, the bearing indication remains displayed to allow the crew to dead-reckon their way back to the landing site in an emergency. Note that this residual bearing is lost when the power is reapplied to the navigation system.

Note – insufficient data is available for bearing computation until the LRV has moved about 50m from the point of 'nav' initialisation, so the display indication should be disregarded until this point on the journey.

Distance indicator
As the name suggests, this instrument displays distance travelled and is calibrated in increments of 0.1km. As described in more detail in Chapter 2, this display is driven from the navigation signal processing unit, which receives its inputs from the third fastest traction drive odometer. Its total digital scale capacity is 99.9km.

In the event of power loss to the navigation system the distance travelled at the time of power loss remains displayed. Re-applying power to the navigation subsystem will cause the indicator to adopt a random number. And operating the system's reset switch will then return the indicator to zero.

Note – driving the LRV in reverse will continue to add to the distance reading. If two wheels are free-wheeling or power is turned off to two wheels, the distance indicated at the time of wheel disengagement or power removal will remain displayed. But driving with two wheels in drag condition (coupled, power off) will give false readings of distance travelled.

Range indicator
This instrument displays the radial distance to the Lunar Module and is graduated in 0.1km increments, with a total scale capacity of 99.9km. In the event of power loss to the navigation system the range indicated at the time of power loss will remain displayed. Reapplying power to the navigation subsystem will cause a random number to be displayed.

Speed indicator
With no speed limits on the Moon, the vehicle's

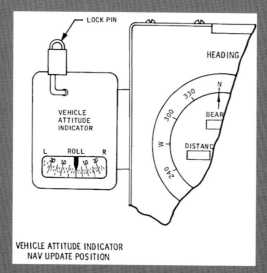

RIGHT Attitude indicator mounted on the left of the Display Console. *(MSFC/NASA)*

speed is not considered all that important. Nevertheless, a speed indicator is still included on the console for interest. The display is driven by the odometer pulses from the right-rear wheel, through the navigation signal processing unit and goes up to 20kmh. Note that when the 'nav' power circuit breaker is open, no speed indication will be attained. And the speed indication will be lost if the right-rear wheel is decoupled or the power is turned off to that wheel.

The Sun shadow device

This device is used to determine the LRV's heading with respect to the Sun azimuth. When deployed, the device casts a shadow on a graduated scale when the vehicle is facing away from the Sun. The point at which the shadow intersects the scale is verbally read to Mission Control for a navigation update. The scale extends 15° either side of zero in divisions of 1°. This device can be utilised at Sun elevation angles of up to 75°.

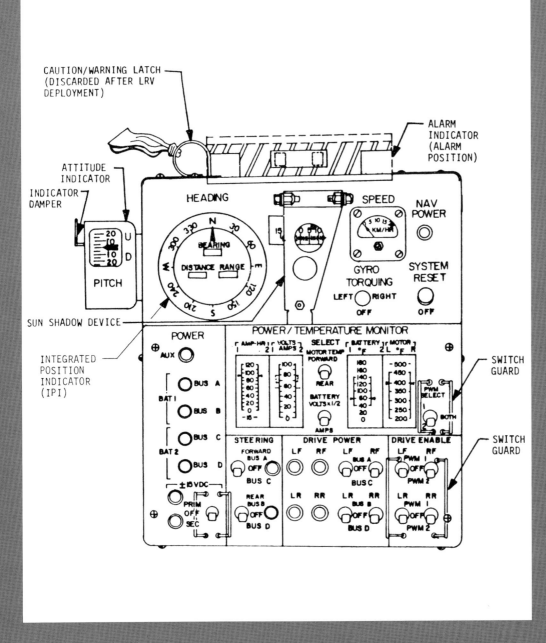

LEFT Layout of the Display Console. *(MSFC/NASA)*

ABOVE David Scott (left) and Jim Irwin (right) practise on an LRV fixed-base simulator in July 1971. (MSFC/NASA)

Navigation initialisation

The navigation system is initialised by momentarily placing the system reset switch to the SYSTEM RESET position and back to OFF. This sets all the digital displays and internal registers to zero. Initialisation should *only* be performed at the start of each EVA in the vicinity of the Lunar Module.

Note that the IPI digital displays reset at approximately 10 counts per second; so to reset the distance indicator from 35.0km to zero (i.e. 350 tenths of a kilometre) takes 35sec.

For full details of how to initialise the navigation system see Chapter 4.

Astronaut training

Astronauts couldn't actually sit in the Lunar Rover's crew station or use the Control and Display Console unless they were on the Moon. The LRV simply wasn't designed to be driven in Earth gravity. So to familiarise themselves with the layout of the crew station and with the Control and Display Console a number of different training methods were developed. These ranged from fixed-base simulators and full-motion simulators to the fully drivable 1G Rover with pneumatic tyres designed for operation on Earth.

As described at the beginning of this chapter, the fixed-base crew station simulators were routinely loaded on board NASA's KC-135 aircraft to fly one-sixth-g parabolas to simulate the lunar environment. Although useful for ergonomic testing, it proved difficult to properly test the vehicle in the cramped conditions on the plane, so most driver training was undertaken on Earth using the specially strengthened 1G Rover which had identical controls.

Astronauts had to be trained to control the

RIGHT Dave Scott (right) and Jim Irwin (left) conduct a suited practice in the 1G Rover. (MSFC/NASA)

ABOVE The one-sixth-g rig suspending the rover. *(MSFC/NASA)*

vehicle precisely with a single gloved hand, sitting inside a bulky space suit with limited visibility. So a simulated lunar surface made from crushed basalt was created near to Boeing's Seattle facility for them to practise on.

To simulate a lunar gravity ride using the 1G Rover, it was hung from a large cable rig with a counterweight system that suspended five-sixths of the vehicle's weight from the ceiling, as it was driven over a simulated lunar surface. Boeing's chief LRV engineer, Gene Cowart, remembers it fondly. "Oh it was amazing, we could get the vehicle to float over the test surface as if it was on the Moon – gave you a sort of giddy feeling to ride – it was so much fun to drive – very useful for the astronauts." Although the 1G Rover had different strength wheels from its lunar counterpart, it gave the astronauts a good idea of the kinds of problems they would face driving on the Moon. "Whenever they hit a bump they actually bounced higher," points out Sam Romano, GM's Rover programme manager. "The steering of course was difficult to duplicate. But we got a good feel for what they were faced with in driving the vehicle in the lunar gravity."

Marshall also built an outdoor Lunar Surface Simulator at KSC in Florida for Rover development and crew training, recreating as closely as possible the soil textures and land contours of the actual lunar surface.

For more 'gritty' fieldwork training in Arizona, the Astrogeology branch of the USGS built a relatively crude, but very useful 1-g rover called the 'Geology Rover' or 'Grover' as the astronaut's dubbed it (see Chapter 1).

Many of the engineers who worked so hard on the Apollo Lunar Roving Vehicle got to try out these various simulators, feeding their childhood dreams of exploring other worlds, but only six men would get to drive the actual vehicles across the surface of another world.

DIFFERENCES WITH THE 1G ROVER

The 1G Rover hand controller operation (speed, steering and brakes) is identical to the LRV hand controller operation, with the following exception: if the hand controller is in full throttle position when the full brakes are applied, drive power will not be automatically cut out.

The 1G Rover seats are also equipped with removable seat pads which allow comfortable operation in a 'shirt sleeve' training session.

The 1G Rover floor panels are flat plates in lieu of beaded panels and are capable of supporting the full weight of standing astronauts in terrestrial gravity.

"Believe it or not, this is like in the training building! The only thing we don't have, Tony, is the linoleum on the floors."

Charlie Duke commenting on the deployment of LRV-2 during the Apollo 16 mission.

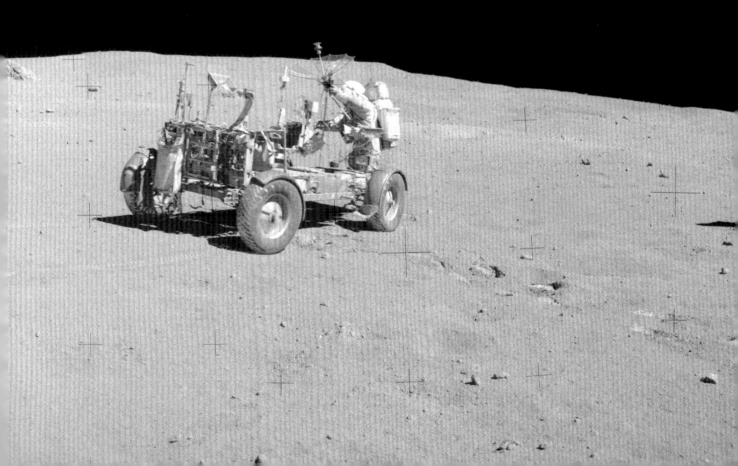

Chapter Six

Stowage and deployment

It is 10 March 1971, and an elated gathering of engineers has assembled at Boeing's Seattle facility to hand over LRV-1 to NASA. It is a big occasion. Although less than 17 months since the inception of the Lunar Roving Vehicle programme, and only 13 months since the prime contract was awarded, it is delivered two weeks ahead of schedule.

OPPOSITE LRV-2 parked at Shadow Rock, Station 13, as John Young aims the high gain antenna at Earth.
(Charlie Duke/NASA/David Woods)

To no one's great surprise, it has cost almost precisely twice Boeing's original estimate, coming in at $38 million. But thanks to careful management on Sonny Morea's part this is right on budget for the figure NASA had given him to produce an Apollo Lunar Roving Vehicle! Boeing hasn't made a single dime from the project, but have delivered a wonder of design and engineering of which everyone is proud. Their fully functional vehicle can be folded up, bolted to the side of a Lunar Module, flown to the Moon, unfolded, and driven by astronauts to places they could never have reached on foot.

All the major NASA managers, and those from Boeing and GM are present. They represent over 800 engineers from NASA and its contractors who have laboured intensely on the project.

"We were extremely happy that we made it just by a day or two ahead of schedule," remembers Sam Romano, GM's Rover project manager. And today, ahead of the vehicle being folded up and lovingly laid in shipping containers for flight to the Kennedy Space Center, they are taking a moment to reflect on their achievements. With launch day just over four months away, all those who worked so hard to create this marvel could only wait to see if their labours and their ingenuity would be rewarded with success.

Stowage

No other wheeled vehicle in history has ever been designed to such exacting standards as the Lunar Rover. What is even more remarkable is that it was designed to be unloaded and assembled in just 15min without tools, jacks or rigs by two men wearing pressurised suits which limited their vision and dexterity.

For this to be possible, the design of the deployment mechanism had pushed the ingenuity of the engineers as hard as any other Rover system. It was fashioned to function as flawlessly as a Swiss watch. But to work properly it still demanded the diligent folding of the Rover in a clean room and then meticulous integration of the vehicle into the deployment mechanism inside Quadrant 1 of the Lunar Module. The vehicle would have to be suspended at all times during this process, because its wheels could not support its own Earth-weight. Every detail of this stowage had to be as precisely conceived and executed as its original design and manufacture had been.

For details of the development story of this stowage and deployment system see Chapter 1.

Once NASA had accepted LRV-1, it was folded, covered, and crated for shipment to the launch site. On arrival at KSC, engineers unfolded the vehicle, checked it out again and conducted another crew station review prior to preparing it for stowage inside the Apollo 15 Lunar Module.

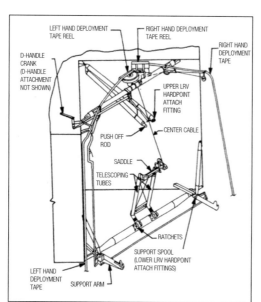
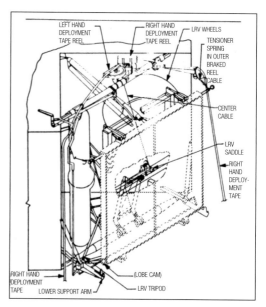

RIGHT Structural Support System and Deployment Hardware. *(NASA)*

RIGHT AND CENTRE Structural Support System and Deployment Hardware, showing tripods highlighted in red. *(NASA)*

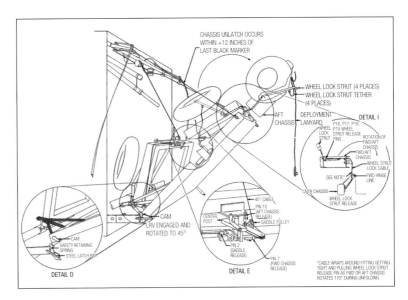

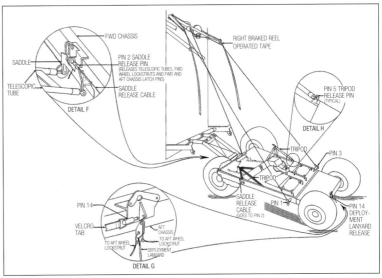

Packing the LRV for the LM wasn't just a matter of folding it up. As Gene Cowart explains, "You had to have a big mechanism that fitted it into the side of the LM." Known as the Space Support Equipment (SSE), this structure inside the LM bay was designed to support the Rover and prevent premature deployment by the violent forces of launch and a potentially hard lunar landing.

The SSE was constructed from two steel support spools that were bolted to Grumman's 'attach fittings' at the bottom of the LM bay, one on each side of the compartment. Aluminium tube tripod structures were clamped to these steel spools and attached to the central section of the chassis of the LRV. Once deployed, their centre legs would double up as outboard toeholds and wheel decoupling tools (see Chapter 5).

The folded vehicle was restrained against rotation about the spool, inside the LM, by a single aluminium strut, which connected the upper inboard LM quadrant corner structure to a so-called 'standoff' point on the LRV centre chassis.

Supported by the SSE structure and its associated hardware, the Rover (without its batteries) would continue to be monitored and

BELOW AND RIGHT LRV-1 being prepared for installation in the Lunar Module at the Kennedy Space Center. Note the stripes on the wheel hubs, used for performance analysis on the Moon. *(NASA)*

would receive its final 'go' for the mission whilst the rest of the Apollo hardware and the Saturn V launch vehicle underwent further checkouts in the Vehicle Assembly Building (VAB). Then the Crawler would carry the entire Saturn/Apollo vehicle the 3½ miles to the launch pad, where it would undergo constant check and recheck right up until launch date.

Finally, with just 18 hours remaining to lift off, the LRV batteries would be installed (see Chapter 3). The next time the vehicle would be checked out was on the Moon!

Deployment

Deployment was an elaborate and carefully devised affair that required bell-cranks, linkages and pins to release the Rover from its structural supports on the LM. To perfectly place the vehicle on the lunar surface required the coordinated action of braked reels, operating tapes, braked reel cables, LRV rotation-initiating push-off springs, deployment cables, telescopic tubes, chassis latches, release pin mechanisms, and rotation-support points.

Deployment mechanism operations

As noted previously, the LRV was attached to the SSE and deployment mechanism and held in position in Quadrant 1 as shown opposite.

Deployment began with the astronauts removing the LM plume heating protection insulation blanket from the outside of the LM and making a visual inspection of the area for obvious signs of damage. It then progressed in five basic phases, noted in the illustration at top right of page 133.

RIGHT Artist's impression of deployment of the Lunar Roving Vehicle onto the lunar surface. *(Grumman)*

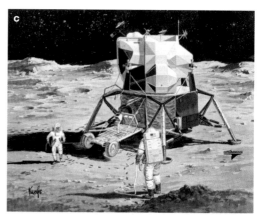

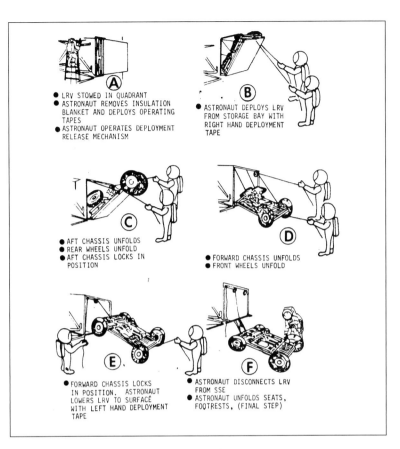

ABOVE LRV-1 in its stowed position on Apollo 15's Lunar Module. *(NASA)*

■ Phase I – Deployment from the stowed position of both braked reel operating tapes and deployment cable (above right, inset A).
■ Phase II – Operating the D-handle to disconnect the LRV from the structural support subsystem above right, inset A).
■ Phase III – Operating the right-hand reel to unfold the LRV and lower the aft chassis wheels to the lunar surface (above, inset B, C and D).

■ Phase IV – Operating the left-hand reel to lower the forward chassis wheels to the lunar surface (above, inset E).
■ Phase V – Disconnecting the SSE from the LRV after all four wheels are on the surface (above, inset F).

Further details of the steps making up these deployment phases and the subsequent starting-up procedures are reproduced from NASA's original operation manual (in American English) spanning the next few pages. Many of these details also appeared on the astronaut's wrist checklists (one page reproduced below).

ABOVE Deployment sequence as documented in NASA's original Lunar Roving Vehicle Operations Handbook under contract NAS8-25145. *(NASA)*

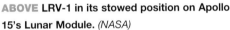

LEFT Cuff checklist for commencement of LRV deployment. *(NASA/Apollo Lunar Surface Journal)*

Rover deployment and start-up instructions

The following tables and associated illustrations, describing the full procedures for Rover deployment and start-up, are taken directly from section 2.1 of NASA's Operations Handbook, which details the normal procedures to be followed by the astronauts for operating the LRV on the lunar surface and the 1G trainer during training operations. The figures referred to in these tables relate to NASA's original figure numbers, also appearing here on the relevant reproduced pages.

The start-up procedures contained in this chapter also apply to the 1G Rover, but not the deployment activities. When performing training, all steps can be performed except those dealing with pulling the LM D-rings, removing the insulation blanket, pulling the deployment tapes, inspecting the hinge pins, and deploying the inboard handholds.

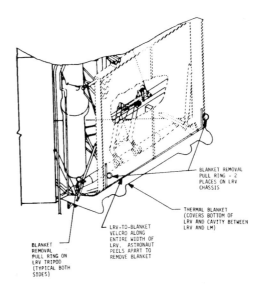

FIGURE 2-1 LRV THERMAL BLANKET

Mission J   Basic Date 12/4/70   Change Date 7/7/71   Page 2-3

STA/STEP	PROCEDURE	REMARKS
2.1	UNLOADING AND CHASSIS DEPLOYMENT Both crewmen are utilized for the unloading. Procedure steps subsequent to 2.1w are for one crewman unless otherwise noted.	Prior to initiation of the subsequent procedures, crew should position TV camera to monitor all LRV deployment operations.
a.	Perform an overall inspection of the LRV and SSE to verify proper configuration and no obvious damage.	To extent which lighting & landing angle permits inspect lower tube & bellcrank assys, left side tube assy, & upper tube & bellcrank assys & verify no obstructions preventing operation of these pin release mechanisms.
b.	Release LRV insulation blanket as follows: (1) At left side pull velcro straps & release blanket from tripod & peel away from floor panel velcro. (2) At right side pull velcro straps & release blanket from tripod & peel away from floor panel velcro. (3) Verify blanket detaches from LRV and falls back toward LM.	Figure 2-1. Perform overall inspection of those portions of LRV and SSE which were previously obscured by insulation blanket.
c.	Inspect <u>each</u> of two lower support arm latches to verify proper configuration. Latch should be in position and trip arm should be up, as determined by continuous white stripe on outer side of lower support arm. If either latch is in the "tripped" position, reset as follows: (1) Push trip arm down. (2) Rotate latch up into position. (3) Raise trip arm until locking dog on trip arm engages receptacle on latch. This will be indicated by continuous white stripe.	Figure 2-1.1.
d.	Release left hand deployment tape stowed in nylon bag attached to lower left support arm by velcro tape.	Figure 2-2.
e.	Stow left hand deployment tape by draping it over a LM landing strut for convenient future access.	Tape should be placed so that crewman is not required to move

Mission J   Basic Date 12/4/70   Change Date 7/7/71   Page 2-2

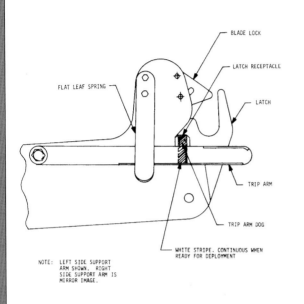

FIGURE 2-1.1 SUPPORT ARM LATCH MECHANISMS LATCHED CONFIGURATION

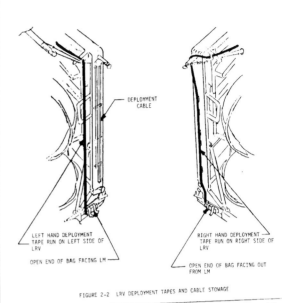

FIGURE 2-2 LRV DEPLOYMENT TAPES AND CABLE STOWAGE

STA/STEP	PROCEDURE	REMARKS
2.1	(Continued)	into limited space between LRV and LM landing leg.
f.	Release deployment cable from teflon clips on left side of LRV center chassis and deploy cable to maximum length and at 45° angle from Quad I toward descent ladder.	
g.	Release right hand deployment tape stowed in nylon bag attached to lower right support arm by velcro tape. Place tape in convenient location for future access.	
h.	Ascend LM ladder.	
i.	Inspect D-handle and bellcranks to verify there are no obstructions preventing operation.	Figure 1-41.
j.	Inspect cable assembly connecting D-handle and bellcrank to ensure there is no fouling that would prevent operation.	
	CAUTION	
	During and subsequent to deployment D-handle operation, both crewmen should remain out of the LRV deployment envelope.	Figure 2-3.
k.	Pull LRV deployment D-handle. Verify LRV moves outward from LM about 4 degrees.	Figure 2-4.
	NOTE: If push-off rod fails to rotate LRV about 4° outward from LM quadrant, deployment cable may be pulled to initiate LRV movement.	First 5 to 6 inches of travel releases lower release pins, lower half of apex fittings may fall away immediately or during deployment rotation. The last segment of travel releases the upper pin. As the upper release pin is pulled, LRV rotates out of LM about 4 degrees.

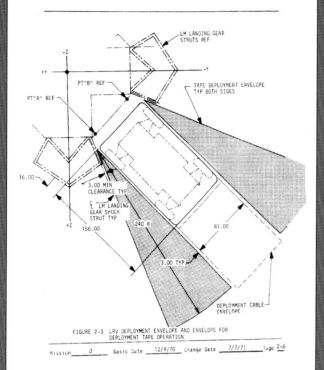

FIGURE 2-3 LRV DEPLOYMENT ENVELOPE AND ENVELOPE FOR DEPLOYMENT TAPE OPERATION

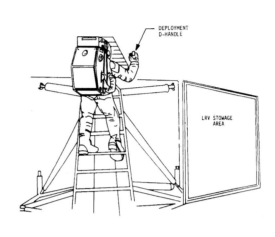

FIGURE 2-4 CREWMAN POSITIONED TO DEPLOY LRV

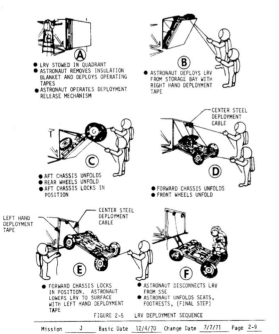

FIGURE 2-5 LRV DEPLOYMENT SEQUENCE

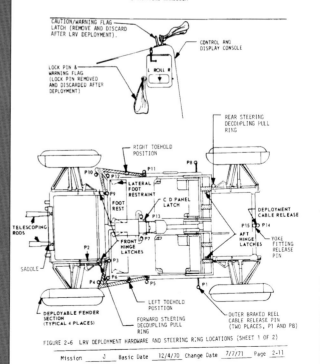

FIGURE 2-6 LRV DEPLOYMENT HARDWARE AND STEERING RING LOCATIONS (SHEET 1 OF 2)

STA/STEP	PROCEDURE	REMARKS
2.1	(Continued)	

l. Descend LM ladder. Grasp deployment cable and monitor deployment activity.

> **CAUTION**
>
> Crewman operating deployment cable should keep slack out of double braked reel cables. Crewman should remain out of LRV deployment envelope but in deployment cable envelope immediately to rear of LRV for direct cable pull.

m. Other crewman pulls right hand deployment tape at side of vehicle. Verify LRV rotates outward from LM (Figure 2-5, View B).

    NOTE: Crewman should remain within defined envelopes for deployment tape operation (Figure 2-3) to ensure that deployment tapes do not contact sensitive LM components.

    Remarks for m: For first 15 degrees of rotation LRV rotates on apex fittings, thereafter apex fittings lift off spools and rotation point shifts to walking hinge. Lower telescopic tubes ratchet engage at 35 degrees rotation.

n. Continue to pull right hand deployment tape (Fig. 2-5, View C). When vehicle rotates outboard to about 45 to 50 degrees, verify that:

    (1) Aft chassis unfolds and locks in position.
    (2) Rear wheels unfold and tethered rear wheel struts fall free.
    (3) Forward chassis is released from console post.

    NOTE: If either aft or forward chassis latch pins fail to pull automatically, deployment cable may be pulled to accomplish pin release.

    Remarks for n: At about 45 degrees the center steel deployment cable tightens, pulling the forward and aft chassis latch pins at the console post mount on the center chassis. The aft chassis and wheels fully deploy and the forward chassis returns to the 35° position. Prior to aft chassis unlatch, the double braked reel operating tape will show black and white stripe markings at the outrigger fairlead as a warning to the crewman.

STA/STEP	PROCEDURE	REMARKS
2.1	(Continued)	

o. Continue to pull right hand deployment tape (Fig. 2-5, View D). Verify that:

    (1) Center/aft chassis rotates until rear wheels contact lunar surface.
    (2) Rear wheels slide on surface permitting center/aft chassis to move away from LM.

    NOTE: If wheels fail to slide, deployment cable may be pulled to permit center/aft chassis to move away from LM.

    Remarks for o: At about 73 degrees, the cam on forward sides of center chassis strikes latch lock arm, forces arm down out of retaining spring and unlocks latch.

p. Continue to pull right hand deployment tape (Fig. 2-5, View D). Verify that:

    (1) Forward chassis continues to unfold and locks in position.
    (2) Forward wheels unfold and tethered front wheel struts fall free.
    (3) Center steel deployment cable again becomes taut.

    Remarks for p: Forward wheel lock strut pins release and forward wheels deploy as the angle between the forward and center chassis approaches 170 degrees. (The center steel deployment cable again becomes taut).

q. Continue to pull right hand deployment tape until outer braked reel cables are slack. Release right hand deployment tape and at chassis RR grasp outer braked reel cable in right hand and remove cable pin P8 (Figure 2-6) with left hand.

    Remarks for q: At this time the forward and aft chassis sections are deployed and locked to the center chassis. All wheels are deployed. The forward chassis is held up by the telescopic tube assembly and the center steel deployment cable.

r. Discard cable and pin outside work area.

STA/STEP	PROCEDURE	REMARKS
2.1	(Continued)	
s.	At chassis LR grasp outer braked reel cable in left hand and remove cable pin P1.	Figure 2-6.
t.	Discard cable and pin outside work area.	
	CAUTION To prevent any sudden surprise LRV motions, the center steel deployment cable should always be under tension when pulling on left hand deployment tape. This is accomplished by maintaining a force on the deployment cable.	
u.	Pull left hand deployment tape (Fig. 2-5, View E). Verify that forward chassis lowers until all wheels contact lunar surface and support vehicle weight and center cable is slack.	This tape was previously stowed over a LM landing strut for convenient access.
	NOTE: If wheels fail to slide, deployment cable may be pulled to move LRV away from LM.	Use of the deployment cable to drag the LRV with the brakes locked shall be kept to a minimum.

Mission J Basic Date 12/4/70 Change Date 7/7/71 Page 2-10.1

LUNAR ROVING VEHICLE
APOLLO OPERATIONS HANDBOOK

STA/STEP	PROCEDURE	REMARKS
2.1	(Continued)	
v.	Release inboard handhold velcro tiedown strap (stand at left side of vehicle to effect release).	Figure 2-9, View 1
w.	Coil deployment cable and remove cable release pin P14 and chassis delatch fitting pin P15. Discard cable and deployment hardware outside of work area.	Figure 2-6. When pin P14 is pulled, deployment cable and rear wheel tethers fall free of vehicle. When pin P15 is pulled, yoke will either fall to surface or be retained in pinless fitting. If retained, pull yoke free at clevis and discard yoke.
x.	Deploy RF fender extension.	
y.	Verify both hinge pins flush at RF hinge.	If hinge pin is not flush, tap pin with toehold subsequently removed in step ac. Verify pin is latched by pressing down on chassis.
z.	Remove pins P9 and P10 from right tripod and discard clear of deployment area.	Figure 2-6.
aa.	Grasp tripod apex with right hand and remove pin P11 with left hand.	
ab.	Discard tripod main members and pin clear of deployment area.	
ac.	Grasp remaining short tripod member in right hand, remove pin P12 with left hand, and discard pin clear of deployment area.	

Mission J Basic Date 12/4/70 Change Date 7/7/71 Page 2-13

LS006-002-2H
LUNAR ROVING VEHICLE
OPERATIONS HANDBOOK

LS006-002-2H
LUNAR ROVING VEHICLE
OPERATIONS HANDBOOK

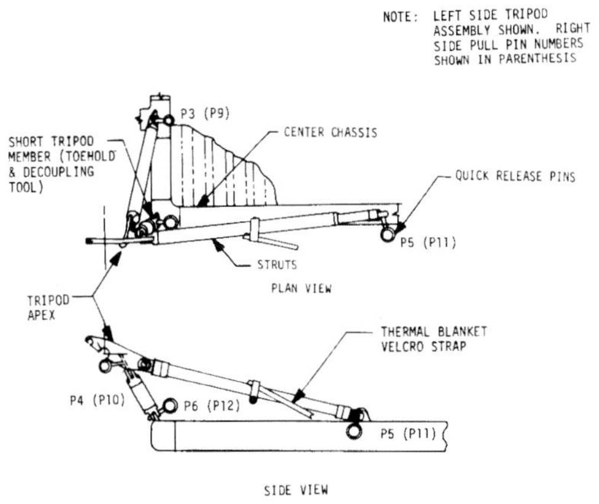

FIGURE 2-6 LRV DEPLOYMENT HARDWARE LOCATIONS (SHEET 2 OF 2)

Mission J Basic Date 12/4/70 Change Date 7/7/71 Page 2-12

LS006-002-2H
LUNAR ROVING VEHICLE
OPERATIONS HANDBOOK

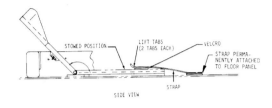

FIGURE 2-7 FOOT REST DEPLOYMENT

Mission J Basic Date 12/4/70 Change Date 7/7/71 Page 2-15

LS006-002-2H
LUNAR ROVING VEHICLE
OPERATIONS HANDBOOK

STA/STEP	PROCEDURE	REMARKS
2.1	(Continued)	
ad.	Remove short tripod member and insert tripod member in right toehold position or stow in underseat stowage bag.	If short tripod member is installed in toehold position, end with hook should be outboard with hook pointing forward. This is also used as wheel decoupling tool.
ae.	Pull right footrest lift tabs.	Figure 2-7. Tabs pull free of footrests but remain attached to the floor panel.
af.	Rotate footrest upward and forward and lock into position.	
ag.	Release velcro tiedown strap (if necessary), pull out right C/D console "T" handle P13 with left hand and turn 90° CW.	Figure 2-8.
ah.	Rotate right seat to stable overcenter position.	
ai.	Rotate legs to full upright position.	
aj.	Attach forward seat legs velcro strap to outboard handhold.	
ak.	Verify underseat stowage bag erects.	
al.	Release right seat belt from underseat bag stowage position and stow in temporary location.	
am.	Pull seat pan frame forward to engage front legs.	
an.	Verify all seat latches latched.	
ao.	Verify both hinge pins flush at RR hinge.	If hinge pin is not flush, tap pin with toehold. Verify pin is latched by pressing down on chassis.

Mission J Basic Date 12/4/70 Change Date 7/7/71 Page 2-14

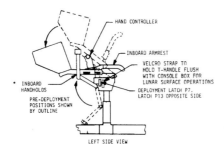

FIGURE 2-8 CONTROL AND DISPLAY CONSOLE DEPLOYMENT

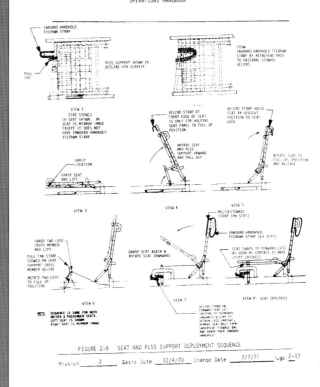

FIGURE 2-9 SEAT AND PLSS SUPPORT DEPLOYMENT SEQUENCE

STA/STEP	PROCEDURE	REMARKS
2.1	(Continued)	on chassis.
ap.	Visually verify the rear steering decoupling pull ring seal has not been broken.	Figure 2-6. If pull ring tie-down seal is broken and subsequent steering check using hand controller indicates steering is engaged, disregard broken seal. If hand controller is not engaged, recouple steering.
aq.	Deploy RR fender extension.	
ar.	Deploy LR fender extension.	
as.	Verify both hinge pins flush at LR hinge.	If hinge pin is not flush, tap pin with toehold. Verify pin is latched by pressing down on chassis.
at.	Rotate left seat to stable overcenter position.	
au.	Rotate legs to full upright position.	
av.	Attach forward seat legs velcro strap to outboard handhold.	
aw.	Verify underseat stowage bag erects.	
ax.	Release seat belt from underseat bag stowage position and place in temporary storage position.	
ay.	Pull seat pan frame forward to engage front legs.	

STA/STEP	PROCEDURE	REMARKS
2.1	(Continued)	
az.	Verify all seat latches latched.	
ba.	Stow inboard handhold tiedown strap by making loop behind seat and attaching end of strap to velcro patch on top of seat back.	
bb.	Fold inboard armrest down.	To prevent interference with hand controller armrest must be folded down to extent possible at this point.
bc.	Support console with left hand, with right hand release velcro tiedown strap (if required), pull out left C/D console "T" handle P7 and turn 90° CW.	
bd.	With right hand rotate inboard handhold to locked position while rotating console downward with left hand.	
be.	Rotate "T" handle P7 90° CW with right hand, fold "T" handle flush with console box and secure in position with velcro strap.	"T" handle should "snap-in", lock and fold down flush with console box.
bf.	Remove attitude indicator lock pin and discard.	Figure 2-6.
bg.	Remove C&W flag lock pin and discard.	
bh.	Pull pins P3 and P4 and discard clear of work area.	
bi.	Grasp tripod apex with left hand and pull pin P5.	
bj.	Discard pins and apex members clear of work area.	
bk.	Grasp short tripod member in left hand and pull pin P6 with right hand, and discard pin clear of deployment area.	

STA/STEP	PROCEDURE	REMARKS
bl.	Remove short tripod member and use hooked end to pull cable P2.	Figure 2-6. Tool hook interfaces with cable area color coded gold. Deflection of cable releases telescoping rods saddle and forward wheel strut tethers.
bm.	Visually verify that telescoping rods saddle falls away from LRV.	
bn.	Either insert short tripod member in left toehold position or stow in underseat stowage bag.	Figure 2-6. If short tripod member is installed in toehold position, end with hook should be outboard with hook pointing forward. This is also used as wheel decoupling tool.
bo.	Pull left footrest lift tabs.	Tabs pull free of footrests, but remain attached to floor panel.
bq.	Rotate footrest upward and forward and lock into position.	
br.	Verify both hinge pins flush at LF hinge.	
bs.	Deploy LF fender extension.	
bt.	Verify battery no. 1 and SPU dust covers closed and secured to velcro patch.	Verify by applying slight lift force on edge of cover.
bu.	Verify the forward steering decoupling pull ring seal has not been broken.	Figure 2-6. If seal is broken and subsequent steering check using hand controller indicates steering is engaged, disregard broken seal. If hand controller check indicates steering is

STA/STEP	PROCEDURE	REMARKS
2.1	(Continued)	not engaged, center wheels in neutral steer, verify forward steering lock and continue mission using rear steering only.
bv.	Move to right side of vehicle and verify battery no. 2 dust cover closed and secured to velcro patch.	Verify by applying slight lift force on edge of cover.
bw.	At right side of LRV rotate right "T" handle P13 90° CW, fold "T" handle flush with console box and secure in position with velcro strap.	"T" handle should snap-in, lock and fold down flush with the console box.

STA/STEP	PROCEDURE	REMARKS
2.2	(Continued)	
	NOTE: The following step (2.2.c) may be performed at crew option. If this step is performed, do not perform steps 2.2.y, 2.2.z, or 2.2.aa.	
c.	Manually move the LRV away from the LM. (See remarks for LRV configuration for this operation).	Crew may manually move LRV away from LM prior to powerup; the hand controller should be placed in neutral throttle position and parking brake released. With a crewman standing on either side of vehicle outboard handholds may be used to lift, move, and tow LRV to and desired location.
	NOTE: After LRV is removed from area adjacent to LM, take precautions to prevent entanglement with wheel lock struts and cables remaining on lunar surface.	

STA/STEP	PROCEDURE	REMARKS

2.2 LRV POST DEPLOYMENT CHECKOUT AND DRIVE TO MESA

a. Verify Hand Controller in parking brake and neutral throttle position and reverse inhibit switch is on (pushed down).

Crewman stands along side the vehicle.

b. Verify switches and circuit breakers in pre-launch positions as follows:

    NAV POWER Circuit Breaker - Open
    GYRO TORQUING Switch - OFF
    System RESET Switch - OFF
    AUX Circuit Breaker - Open
    BUS A, B, C, D, Circuit Breakers - Open
    $\pm$ 15 VDC PRIM and SEC Circuit Breakers - Open
    $\pm$ 15 VDC Switch - OFF
    MOTOR TEMP Switch - FORWARD
    BATTERY Switch - AMPS
    PWM SELECT Switch - BOTH
    STEERING FORWARD and REAR Circuit Breakers - Open
    STEERING FORWARD and REAR Switches - OFF
    DRIVE POWER LF, RF, LR, RR Circuit Breakers - Open
    DRIVE POWER LF, RF, LR, RR Switches - OFF
    DRIVE ENABLE LF, RF, LR, RR Switches - OFF
    Verify all meters, indicators are at zero. Report off-zero indications.

Figure 2-11. Crewman stands along side vehicle.

2.2 (Continued)

Lunar weight of LRV at this point would be approximately 85 lbs. Hand controller is placed in neutral throttle position and brake disengaged to permit wheels to roll.

d. Set parking brake. (Or if step 2.2.c was not performed, verify brake is set).

Crewman stands along side vehicle, and should exercise care not to move vehicle while setting brake.

e. Ingress left seat, fasten seat belt and initiate subsequent power up steps.

Figure 2-10.

f. BUS A, BUS B, BUS C, BUS D Circuit Breakers - Close.

g. BATTERY Switch - VOLTS x 1/2.

h. Report BAT 1 and BAT 2 VOLTS indications.

i. BATTERY Switch - AMPS.

j. Report BAT 1 and BAT 2 temp (°F) indications.

k. Report BAT 1 and BAT 2 AMP-HR indications.

l. Report BAT 1 and BAT 2 AMPS indications.

m. $\pm$ 15 VDC PRIM and SEC Circuit Breakers - Close.

n. STEERING FORWARD AND REAR Circuit Breakers - Close.

o. DRIVE POWER LF, RF, LR, RR Circuit Breakers - Close.

LS006-002-2H
LUNAR ROVING VEHICLE
OPERATIONS HANDBOOK

SEAT BELT PLACED ON INBOARD HANDHOLD FOR STORAGE PRIOR TO VEHICLE EGRESS

FIGURE 2-10 CREW POSITION

Mission ___J___ Basic Date _12/4/70_ Change Date _7/7/71_ Page _2-24_

LS006-002-2H
LUNAR ROVING VEHICLE
OPERATIONS HANDBOOK

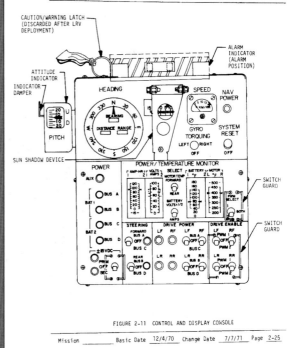

FIGURE 2-11 CONTROL AND DISPLAY CONSOLE

Mission _____ Basic Date _12/4/70_ Change Date _7/7/71_ Page _2-25_

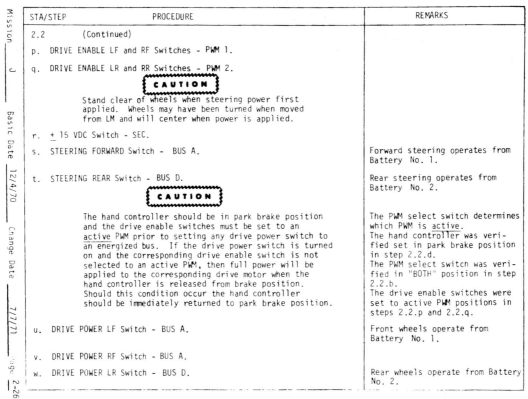

STA/STEP	PROCEDURE	REMARKS
2.2	(Continued)	
p.	DRIVE ENABLE LF and RF Switches - PWM 1.	
q.	DRIVE ENABLE LR and RR Switches - PWM 2.	

CAUTION

Stand clear of wheels when steering power first applied. Wheels may have been turned when moved from LM and will center when power is applied.

r.	± 15 VDC Switch - SEC.	
s.	STEERING FORWARD Switch - BUS A.	Forward steering operates from Battery No. 1.
t.	STEERING REAR Switch - BUS D.	Rear steering operates from Battery No. 2.

CAUTION

The hand controller should be in park brake position and the drive enable switches must be set to an <u>active</u> PWM prior to setting any drive power switch to an energized bus. If the drive power switch is turned on and the corresponding drive enable switch is not selected to an active PWM, then full power will be applied to the corresponding drive motor when the hand controller is released from brake position. Should this condition occur the hand controller should be immediately returned to park brake position.

		The PWM select switch determines which PWM is <u>active</u>. The hand controller was verified set in park brake position in step 2.2.d. The PWM select switch was verified in "BOTH" position in step 2.2.b. The drive enable switches were set to active PWM positions in steps 2.2.p and 2.2.q.
u.	DRIVE POWER LF Switch - BUS A.	Front wheels operate from Battery No. 1.
v.	DRIVE POWER RF Switch - BUS A.	
w.	DRIVE POWER LR Switch - BUS D.	Rear wheels operate from Battery No. 2.

Mission ___J___ Basic Date _12/4/70_ Change Date _7/7/71_ Page _2-26_

LUNAR ROVING VEHICLE
APOLLO OPERATIONS HANDBOOK

STA/STEP	PROCEDURE	REMARKS
2.2 (Continued)		

x. DRIVE POWER RR Switch - BUS D.

NOTE: Do not perform the following steps 2.2.y, 2.2.z, and 2.2.aa if step 2.2.c was performed.

y. Release Parking Brake

z. Hand Controller reverse inhibit switch - UP position.

   NOTE: The LRV driver may now back away from LM. LRV driver should request other crewman to direct and monitor any backing operations from an off-vehicle position.

aa. Stop LRV and set parking brake. Reset reverse inhibit switch (push switch down).

ab. Release parking brake and drive to MESA area for equipment loading.

    Remarks: To the extent possible driver should verify steering, speed control and braking during this brief drive. The off-vehicle crewman should verify all four wheels rotating (not sliding).

ac. Stop LRV and set hand controller in parking brake position; Neutral throttle.

    Remarks: Parking brake should always be set prior to vehicle egress by either crewman.

ad. Perform LRV partial power down as follows:

    Remarks: Turning off drive power, steering, and + 15 VDC switches ensures that a failure in the DCE will not apply power to any vehicle motor thereby precluding any unnecessary power drain.

STA/STEP	PROCEDURE	REMARKS
2.2 (Continued)		

ad. (Continued)

    DRIVE POWER Switches (4) - OFF.
    STEERING Switches (2) - OFF.
    + 15 VDC Switch - OFF.

   NOTE: The above step 2.2.ad assumes that payload loading and first LRV traverse will follow in order. Should crew rest period be scheduled subsequent to step 2.2.ae and prior to first LRV sortie, then Bus A, B, C, and D circuit breakers should be opened.

ae. Release and stow seat belt and egress vehicle.

STA/STEP	PROCEDURE	REMARKS
2.3	PAYLOAD LOADING	
2.3.1	LCRU Installation	
a.	Place LCRU support post locks in the up position.	Figure 2-12. LRV arrives on lunar surface with LCRU support posts installed in LRV support tubes on forward chassis and with LRV/LCRU power cable connected to LRV auxiliary connector.
b.	Disconnect GCTA connector from LRV dummy connectors.	Figure 2-12.
	NOTES	
	1. Do not disconnect LCRU power cable from LRV auxiliary connector. Dust contamination could occur if this connector is disconnected.	
	2. Do not allow GCTA connector of cable to fall to lunar surface.	
	3. Do not place payload on battery cover.	
c.	Remove dummy connector from LRV GCTA receptacle and discard.	
d.	Remove LCRU from its LM stowage position and place onto LRV forward chassis LCRU support posts.	Figure 2-13.
e.	When LCRU is bottomed against support posts, position support post locks in horizontal position to secure LCRU.	
f.	Verify LRV AUX power circuit breaker - Open.	

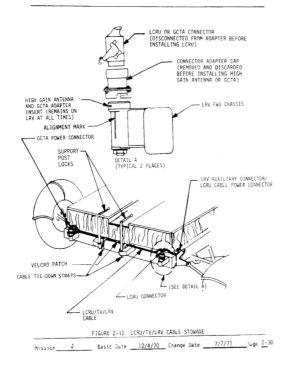

FIGURE 2-12. LCRU/TV/LRV CABLE STOWAGE

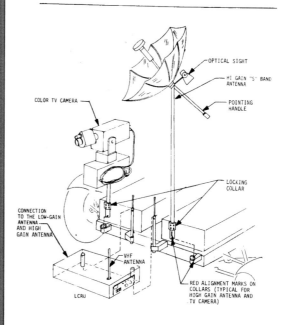

FIGURE 2-13. LCRU, HIGH GAIN ANTENNA, TV CAMERA INSTALLATION

STA/STEP	PROCEDURE	REMARKS
2.3.1	(Continued)	
g.	Disconnect LCRU power connector from LRV dummy connector and connect to LCRU.	
h.	Cover connector with thermal boot.	
i.	Remove dummy connector from LRV HGA receptacle and discard.	
2.3.2	GCTA Installation	Figure 2-13.
a.	At MESA, pull GCTA control unit pip pin release cable.	
	CAUTION Do not strike GCTA control unit mirror surfaces on MESA.	
b.	Remove GCTA control unit and support staff from MESA.	
c.	Unfold GCTA support staff. Verify staff locked.	
	CAUTION If GCTA staff is not properly locked, it could fall on LCRU and cause severe LCRU radiator damage.	
d.	With connector receptacles inboard, insert GCTA staff into mounting receptacle on right front corner of LRV.	
e.	Rotate staff to assure engagement of staff anti-rotational pins.	
f.	GCTA staff bayonet collar - Lock (CW).	Alignment marks are provided on GCTA staff locking collar.

STA/STEP	PROCEDURE	REMARKS
2.3.2	(Continued)	
g.	Connect GCTA connector of LRV/LCRU cable to GCTA control unit.	
h.	On TV camera, LM PWR Switch - OFF.	
i.	Disconnect LM/TV cable from TV camera and rest connector on tripod handle.	
j.	Remove TV camera from LM tripod and install on GCTA azimuth/elevation unit.	
k.	Connect GCTA control unit connector to TV camera.	
2.3.3	16 mm Data Acquisition Camera Installation	Figure 2-14.
a.	Remove camera and staff from LM.	
b.	Assemble camera and staff into single unit.	
c.	Insert staff into receptacle on LRV right inboard handhold.	
d.	Verify staff locked in place by pulling up on the camera without depressing the push button on end of handhold. Camera staff should not move vertically.	
2.3.4	Low Gain Antenna Installation	Figure 2-14.
a.	Remove low gain antenna from LM stowage location.	
b.	Insert low gain antenna staff on LRV left inboard handhold.	
c.	Verify staff locked in vertically by pulling up on staff without depressing button on end of handhold. Low gain antenna staff should not move vertically.	

STA/STEP	PROCEDURE	REMARKS
2.3.4	(Continued)	
	d. Route low gain antenna cable to LCRU and secure to LRV with strap on console and clips on forward chassis.	Figure 2-15.
	e. Connect low gain antenna cable to LCRU.	
2.3.5	High Gain Antenna Installation.	Figure 2-13.
	a. Remove high gain antenna from LM stowage position.	
	b. Insert high gain antenna staff into the mounting receptacle on the left front corner of the LRV and lock.	Alignment marks are provided on HGA staff locking collar.
	c. Unfold and lock HGA staff.	
	d. Remove and discard optical sight retaining clamp.	
	e. Open and lock HGA dish.	
	f. Connect HGA cable to the LCRU.	
	g. Deploy LCRU VHF whip antenna.	
	h. Activate LCRU/GCTA and perform communication checks as required.	
	i. Deactivate LCRU/GCTA until needed.	
2.3.6	Aft Payload Pallet Installation	
	a. Release the pallet support post tiedown on LRV aft chassis.	Figure 2-16.
	b. Erect pallet support post.	
	c. Remove pallet from LM.	

Mission J Basic Date 12/4/70 Change Date 7/7/71 Page 2-35

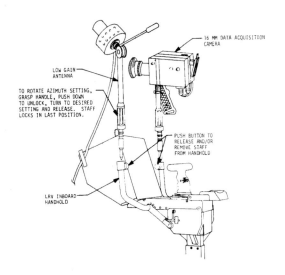

FIGURE 2-14 16 MM DAC AND LOW GAIN ANTENNA INSTALLATION

Mission J Basic Date 12/4/70 Change Date 7/7/71 Page 2-34

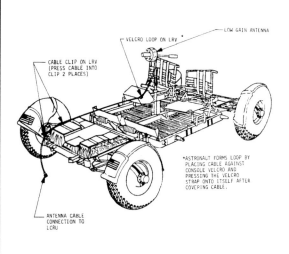

FIGURE 2-15 LCRU LOW GAIN ANTENNA CABLE INSTALLATION ON LUNAR SURFACE

Mission J Basic Date 12/4/70 Change Date 7/7/71 Page 2-36

STA/STEP	PROCEDURE	REMARKS
2.3.6	(Continued)	
	CAUTION	
	LRV steering recoupling tool must be cleared when installing pallet.	
d.	Connect pallet to pallet support post.	
e.	Rotate pallet about support post until pallet locks in pallet adapter on LRV LH aft chassis.	Figure 2-17.
2.3.7	Buddy SLSS Installation	
a.	Remove BSLSS bag from LM.	
b.	Release BSLSS support strap on back of right seat.	
c.	Feed strap through BSLSS bag handle and secure to PLSS support velcro on front of back seat.	Figure 2-18.
2.3.8	Map Holder Installation	
a.	Remove map holder from LM.	Figure 2-18.1
b.	Place map holder clamp around inboard handhold vertical member.	
c.	Tighten clamp by rotating clamp handle CW.	
2.3.9	Gnomon Bag Installation	
a.	Remove gnomon bag from LRV LH underseat stowage bag.	Figure 2-18.2
b.	Secure gnomon bag to back of LH seat per Figure 2-18.2	

STA/STEP	PROCEDURE	REMARKS
2.3.10	Apollo Lunar Surface Drill (ALSD) Installation	Figure 2-18.3
a.	Verify RH seat belt has been removed from RH underseat stowage bag and is stowed over inboard handhold.	The ALSD is placed on the LRV only for transport from the LM to the ALSEP site with only one astronaut on the LRV.
b.	Leave LRV RH seat support in stowed-for-launch position and raise seat pan and PLSS support to configuration shown in View 5, Figure 2-9.	
c.	Remove ALSD from LM.	
d.	Place ALSD over stowed seat support frame and RH underseat bag. Orient ALSD per Figure 2-18.3	
e.	Slide ALSD rearward until ALSD butts against LRV center chassis rear cross-member.	
f.	Lower RH seat pan to operational position.	
	NOTE: Forward seat pan member will rest on ALSD structure. This is satisfactory configuration.	
2.3.11	Laser Ranging Retro Reflector (LR^3) Installation	Figure 2-18.3
a.	Remove LR^3 from LM.	The LR^3 is placed on the LRV only for transport from the LM to the ALSEP site.
b.	Place LR^3 on RH seat of LRV per Figure 2-18.3	
c.	Thread LRV RH seat belt through LR^3 handle and secure seat belt hook to LRV outboard handhold.	

LS006-002-2H
LUNAR ROVING VEHICLE
OPERATIONS HANDBOOK

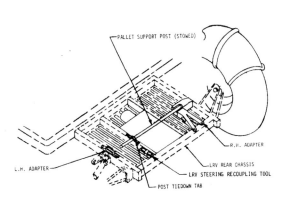

FIGURE 2-16 LRV REAR PAYLOAD PALLET ADAPTERS

Mission J Basic Date 12/4/70 Change Date 7/7/71 Page 2-37

LS006-002-2H
LUNAR ROVING VEHICLE
OPERATIONS HANDBOOK

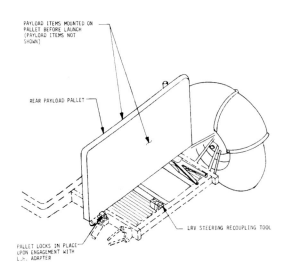

FIGURE 2-17 REAR PAYLOAD PALLET INSTALLED

Mission J Basic Date 12/4/70 Change Date 7/7/71 Page 2-39

LS006-002-2H
LUNAR ROVING VEHICLE
OPERATIONS HANDBOOK

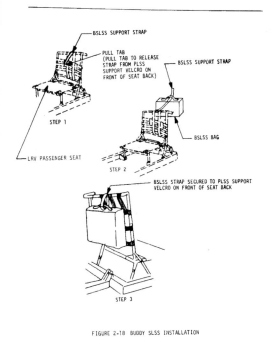

FIGURE 2-18 BUDDY SLSS INSTALLATION

Mission J Basic Date 12/4/70 Change Date 7/7/71 Page 2-40

LS006-002-2H
LUNAR ROVING VEHICLE
OPERATIONS HANDBOOK

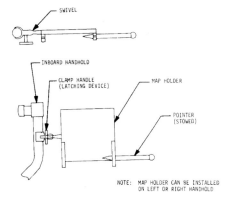

NOTE: MAP HOLDER CAN BE INSTALLED ON LEFT OR RIGHT HANDHOLD

FIGURE 2-18.1 MAP HOLDER INSTALLATION

Mission Basic Date 12/4/70 Change Date 7/7/71 Page 2-40.1

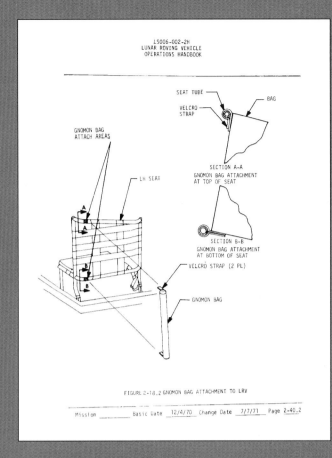
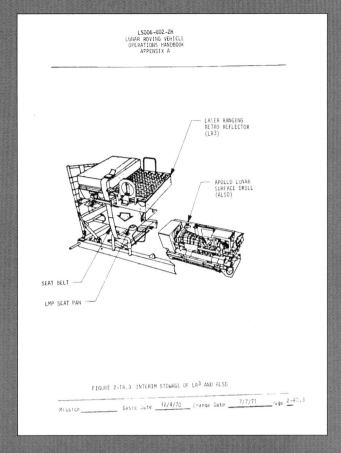

The Rover was deployed facing towards the LM and was turned around by the crew, lifting it using handles on the central chassis. It was driven around the LM and parked adjacent to the MESA in order to commence equipment loading and installation of the LCRU, TV camera and high-gain antenna. Finally, a pre-sortie checkout and preparation was undertaken.

To the Moon

Just after 8.30 am on 26 July 1971, Apollo 15 launched from Pad 39A at the Kennedy Space Center with LRV-1 tucked inside the Lunar Module *Falcon* near the top of the Saturn V stack. Sam Romano, GM's chief Rover engineer, watched from a bleacher as the immense rocket, the heaviest yet to fly, slowly climbed skyward. "I was at the launch, it was an amazing day. After ten years of rover development for me I was finally watching my machine leave for the Moon." His counterpart at Boeing, Gene Cowart, was also there. "At last it was on its way," he reminisces. "But I didn't really believe it until I heard they were go for TLI – for Translunar Injection, the burn that put them on course for the Moon."

Everyone concerned would have to wait another three days before the culmination of their efforts was realised, when the Rover deployment procedure would be put to its first test on the lunar surface.

fast. It wasn't made to go very fast, but it could sure save a lot of energy. You could pack a lot of gear and save a lot of walking time, and it allowed us to cover territory and ground we never would have covered otherwise."

Gene Cernan, Apollo 17 commander

Chapter Seven

Wheels on the Moon

The LRV's performance and driving experience

It is the last day of July 1971. Across America the LRV teams at Boeing, General Motors and the Marshall Space Flight Center are holding their breath. Their collective efforts during the past 17 punishing months will be put to the test in the next few hours. For many of them, today is also the culmination of an entire decade of design, ingenuity, experimentation, prototyping and testing. And for Boeing there's $38 million as well as its reputation riding on the day ahead!

OPPOSITE Harrison 'Jack' Schmitt works on a science experiment near Apollo 17's landing site at Taurus-Littrow. The Lunar Module *Challenger* is in the distance and LRV-3's TV camera is watching him. It is the end of their first day of exploration. *(Gene Cernan/NASA/David Woods)*

Deployment on the Moon

David Scott and Jim Irwin have landed on Hadley Plain, a spectacular embayment between two huge mountains and a meandering canyon, and they are ready for "exploration at its greatest" – as Scott describes it in his 'first step' speech. But they have to get LRV-1 onto the surface and operational to find out if all the hard work, long days and high stress of the Rover project will pay off.

"Let's take a look at our Rover friend here." Scott carefully scrutinises the folded vehicle to check that nothing obvious would interfere with deployment. Irwin positions himself at the top of the ladder to release a latch that holds the Rover in place. Everyone involved in the Rover at Boeing and GM sits forward in their seats; on tenterhooks as they watch the live TV transmission from Hadley.

Then Scott spots something. "Both walking hinges were open, Joe," he tells Capcom Joe Allen in Houston. The hinges at the bottom of the Rover that support it during deployment have become unlatched. Scott quickly resets them before Irwin attempts to release the mechanism. "Okay, here it comes."

For the next few minutes, Scott pulls on a tape that gradually brings the Rover out of the side of the LM. When it gets to 45°, the rear chassis suddenly folds out and its wheels pop into place. Scott continues to pull on the deployment tape until they reach the surface. Then with a little tug from Irwin, the front chassis and wheels fold out to bring the vehicle to its full length. Now only

Ten frames from TV coverage of the deployment of LRV-1 at Hadley-Apennine. In frame 2, Irwin is at the top of the ladder to release the rear of the Rover. By frame 4, the rear chassis has unfolded and its wheels have sprung into position. By frame 6, the rear wheels have reached the ground and the forward chassis is still parked in the LM. By frame 8, the front wheels have deployed, and in the last frame, Scott and Irwin manually lift LRV-1 away from the LM. *(NASA)*

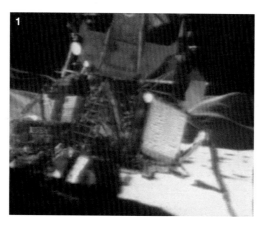

a mechanism with two telescoping arms, known as the saddle, attaches the Rover to the LM. The two astronauts manhandle the car away slightly from the spacecraft.

"Okay. It looks like it's loose to me!" Scott is pleased with progress but he has to finally free it. Although he has released two pins that engage it to the saddle, it is not coming away. The two astronauts then stand on either side and lift the vehicle up and away, but it remains attached to the saddle. The LM's 10° tilt and the uneven ground that the Rover has come down onto are causing the saddle to bind. Romano and Pavlics know that they've tested their deployment mechanism on tilts that are greater than this, but not under exactly these conditions on the Moon before! The tension in the room is vivid. "That was a real problem," remembers NASA engineer Skeet Vaughan. "We were worried about egg on our face if we couldn't get that thing out."

Scott and Irwin have to hold the Rover up in a way that relieves the stress on the saddle.

"There we go," calls Irwin as they find the sweet spot where the Rover can come free.

"Good show," says Scott. "Okay, Joe, it's off." LRV-1 has all four wheels on the Moon! It's a moment that brings tears to Ferenc Pavlics' eyes.

"Outstanding," exults Joe Allen in Houston. The deployment has kept a lot of people awake at night with worry and its successful completion is a major milestone for the team and for the little car.

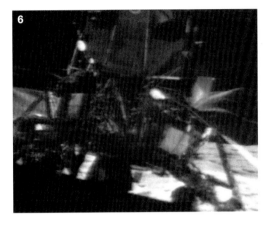

Starting the car

The LRV had reached the Moon, but had its intricate electrical systems survived that aggressive Saturn V launch four days before, the flight to the Moon, and the lunar landing? The world would soon find out. Scott climbed aboard his new machine, and after a struggle with his seat belt he started the process to bring it to life. "Okay, hand controller is locked. Brake's on, reverse is down." He then pushed in all the necessary circuit breakers before telling Mission Control the state of the meters.

"Volts: on number 1 – I've got about 82," which after being divided by two, meant 41V, "and number 2 is reading zero. Hmm." Had they lost a battery already? Scott routed power to the drive and steering systems, took hold of the T-handle and eased it forward from its central position. "Out of detent; we're moving." On the TV, Earthbound engineers and geologists collectively let their breath out as they saw the Rover slip smoothly out of frame.

BELOW Apollo 15 LMP Jim Irwin at LRV-1 on the first day, with a shadowy Mount Hadley in the distance. *(David Scott/NASA)*

It worked! Thoughts of long, tiring walks and being unable to reach hoped-for sites were dismissed. Both batteries were good and only a faulty meter had temporarily fooled them. But another gremlin lay in wait.

"Hey, Jim, you can probably tell me if I've got any rear steering." In his suit, Scott could not turn to see the back of the Rover.

"Yeah, you have rear steering," confirmed Irwin.

"But I don't have any front steering."

This was a big blow for Boeing and GM, as Sam Romano remembered. "I was sitting in the Mission Control Center in the third row. Dr von Braun was in the fifth row. So when they said the front wheels are not steering, I was very, very nervous. The back of my neck began to swell...to get red. My ears were red. And er...we were terrified."

Management's immediate concern was that it could jeopardise plans for the next three days. In the backroom at Mission Control, Pavlics and the other Rover engineers swung into action quickly to troubleshoot the problem while Scott double-checked the Rover's switches. Pavlics suspected a problem in the hand controller, but time was precious and the day's exploration could still be attempted using the rear steering. Capcom Allen quickly radioed up the command, "Press on."

"Okay. That's a good idea," agreed Scott.

In a last ditch effort to fix it before Scott drove off, Allen made a suggestion to help diagnose the issue.

"You might physically try to turn the front wheel; if you think now is a good time."

Scott moved around to the front of the Rover and grasped a wheel with his hands to turn the steering axis. "Aghh! I don't get much out of turning the front wheels."

So without a solution, Scott and Irwin set off on the very first Rover drive on the Moon using rear steering only.

"Going to miss that double Ackermann," rued Scott, referring to the LRV's impressive manoeuvrability with its combined front and rear steering.

On this first drive, 3km to Elbow Crater, the astronauts noted how the front wheels stayed centred. It seemed that something was mechanically locking up the system.

Navigation system performance

At the start of each working day, one of the first jobs for the commander was to set the navigation system up so that it could accurately keep track of their drive and enable them to efficiently return to the Lunar Module.

As the drive progressed, it was Marshall environmental test engineer Skeet Vaughan's job to keep track of how accurately the navigation system was working. "We did a panorama of each site [with the TV camera] and from that we would figure out how much they had travelled according to the navigation system and then we would determine where we thought they were," he explains. The point of this exercise was to collect some real data on just how much wheel slip might be occurring, so they could improve the navigation system for the next mission.

The ultimate accuracy of the navigation system could be gauged by noting whether, upon return to the LM, it agreed with the fact that the Rover had returned to the initialisation point. At the end of their third drive on Apollo 15, a 5.1km visit to the edge of Hadley Rille, Scott parked up and Irwin read out the numbers from the instrument panel. "The heading is 001, 032, 5.1, 0.0." The second number, their bearing to the LM, was meaningless because near to their initialisation point, the reading becomes ambiguous. Of more interest was the last figure – the zero! As far as the navigation system was concerned, they were within 100m of their start, which was very likely true.

"Copy, Jim. Remarkable nav system." Allen spoke for all the engineers who had worked on it.

"Sure is," agreed Irwin.

Off-road driving

As soon as he let the Rover loose in its home environment, Scott was keen to pass on his impressions to Houston for the benefit of the teams of LRV engineers listening across the US – and for the wider world – curious about what it felt like to drive on the Moon! "Okay, Joe, the Rover handles quite well. We're moving at, I guess, an average of about 8 kilometres an hour." Its speed was up to expectations. Scott continued, "It's got very

ABOVE LRV-1's final expedition was to the edge of Hadley Rille. The winding canyon is behind the Rover where David Scott is working. St George Crater and Mount Hadley Delta are beyond in this image stitched from photographs taken by Jim Irwin. *(Jim Irwin/NASA/David Woods)*

BELOW The Earth-facing hemisphere of the Moon indicating the six Apollo landing sites and the major lunar features. Only Apollo 15, 16 and 17 carried a Rover. *(David Woods)*

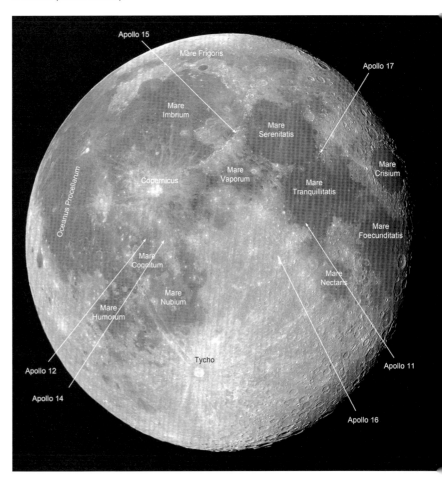

low damping compared to the 1G Rover, but the stability is about the same. It negotiates small craters quite well, although there's a lot of roll. It feels like we need the seat belts, doesn't it, Jim?"

"Yeah, really do," agreed Irwin seated on the right. They had found the seatbelts to be more difficult to fasten than expected because the suit had expanded more in the Moon's very good vacuum and the light gravity meant they sat slightly higher in their seats.

Scott took a glance at how the wheels were coping. Their wire mesh was designed to open up after being compressed against the ground on each rotation. As they did, they let out any soil that had got into the wheel. "There's no accumulation of dirt in the wire wheels."

Allen responded, "Just like in the owners' manual, Dave."

But even at only 8kmh, or 8 'clicks' as the astronauts liked to say, Scott found he had to concentrate hard to drive across the battered lunar surface. He had to keep his eyes forward and was rarely able to look at the Control and Display panel. Scott was a top-class pilot and used to scanning instruments, but in an aeroplane, you don't have a constant parade of boulders and craters large and small coming towards you. He had to keep focused and was glad that Irwin was seated on his right, ready to read out any numbers for him.

"The steering is quite responsive even with only the rear steering," Scott noted on their first day out. "There doesn't seem to be too much slip. I can manoeuvre pretty well with the thing. If I need to make a turn sharply, why, it responds quite well." It turned out to be almost too responsive even with just rear-wheel steering.

"Hang on," hollered Scott as they started down a small slope. "Whoa! Hang on!"

The Rover's lightly loaded back end lost its grip while the front wheels, weighed down with batteries, electronics and the LCRU, dug into the soft soil. In a moment, Scott and Irwin found themselves looking back uphill from where they came as their vehicle did a '180'. "Got to go easy downhill, huh?" laughed the commander.

"I'd say so," agreed Irwin, almost giggling.

"You can't go fast downhill in this thing," explained Scott, "because if you try and turn with the front wheels locked up like that, they dig in and the rear end breaks away, and around you go."

Next morning, Scott tried the front steering again and got a pleasant surprise. "You know what I bet you did last night, Joe? You let some of those Marshall guys come up here and fix it, didn't you?"

Joe Allen in Houston conceded that a lot of people from the Marshall Space Flight Center had been thinking about the problem. "They've been working. That's for sure."

Irwin laughed when he saw what was happening. "It works, Dave?"

"Yes, sir," said Scott. "It's working, my friend. Boeing has a secret booster somewhere to take care of their Rover!"

Most likely, the deep cold of space had caused the steering mechanism to bind, especially during the time since they landed when the Rover, in its stowage bay, was on the shaded side of the LM for their first sleep on the surface. Then a few hours out in the warmth of the Sun during their second sleep had allowed it to free up.

Across three days with the Rover, Scott took the chance to give other steering modes a workout. On his second day, he found the double Ackermann steering almost too sensitive. "The steering is a new task, Joe. It's really responsive now. This thing really turns!"

"Roger, Dave," replied Allen. "We don't want it to be too easy for you."

For a while, he disabled their rear steering. However, he elected not to lock it in its central position in view of the problem they had already had with the front system. "It's really a lot better," he reported. "The double Ackermann's a little too responsive when you have the lack of traction, especially on the slopes." Before long, the rear-wheel steering began to wander, causing the Rover to crab and Scott was forced to resume the double Ackermann mode.

The longest single journey taken by any Rover, 20km, was on the second day out for the Apollo 17 astronauts and Gene Cernan was enjoying the trip. "Let me tell you, this is quite a Rover ride."

"It sure sounds like it," said Bob Parker in Houston.

"But it's quite a machine, I tell you!" enthused Cernan. "I think it would do a lot more than we'd let it."

"The Rover had great manoeuvrability, particularly with the fore and aft steering," recalled Cernan years later. "It's a sporty little ride. It was fun. I don't know how fast you would really want to go up there, quite frankly. You can't see what's up ahead of you too well. And the next thing you know, wow, you're in a hole or you're up [off the ground]."

His colleague Jack Schmitt added to the point. "I'm quite certain that, even at 10 kilometres per hour, there were times when we had all four wheels off the ground. That really made it sporty."

Hill climbs

The Rover was designed as an off-road vehicle from the start, but what surprised many, including its drivers, was just how easily it negotiated uphill climbs, as Scott found on his first drive. "Oh, look at this baby climb the hill," he remarked as he and Irwin headed past Elbow Crater on the lower slopes of Mount Hadley Delta.

"Yeah, climbing at about 8 clicks," noted Irwin.

Immediately, engineers wanted to know how much juice the climb was pulling from the batteries. "It's 10 on batt[ery] 1, Joe," said Irwin.

Each motor was rated at one-quarter horsepower, or 186W. At full power, the current to a motor should be the power divided by the voltage, which for a 36V supply was 5.2A. Since battery 1 was supplying two motors as well as some electronics kit, 10A total was a pretty good figure.

One of the curious aspects of driving on the Moon was that it was sometimes difficult to tell when the Rover was climbing at all, as Scott recalled. "We ended up going up slopes that we don't realise were there, because there's not much force. You're not leaning back or anything in one-sixth [gravity]. You don't notice the lean. And you're focused on the direction and there's not a real horizon around. There are no vertical trees. As far as it feels, you're on a flat surface, going level, when, in fact, you're going up pretty steep."

The Taurus-Littrow landscape, where Apollo 17 landed, included a scarp. As the Moon cooled over its long history, it shrank and in places, this caused the crust to fracture and one side to rise up a little in what geologists call

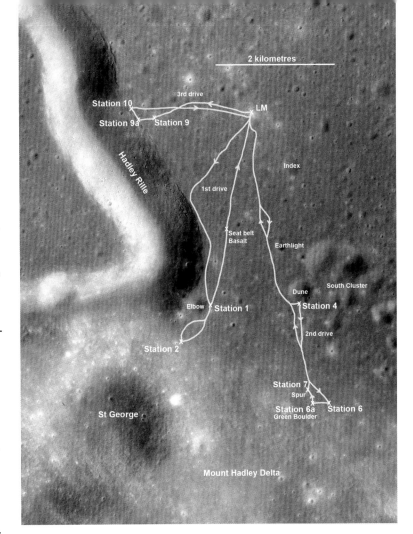

a thrust fault. Cernan and Schmitt had to climb one of these scarps in LRV-3 during their long second-day drive. "I don't even think the Rover knows it's going uphill," reported Cernan.

"You're making about 8 clicks," informed Schmitt.

"And I'm full bore." Cernan had the handle fully forward as he took their Rover to the top of the scarp. "Well, I'll tell you. This Rover doesn't know it's going up a hill."

Even on the slope, Cernan had to continue to dodge obstacles, which meant taking parts of the scarp cross-slope instead of directly upslope. "Get around this crater," he muttered.

"Pretty healthy roll we're going to have here," warned Schmitt.

"Yeah, I'm going to head more straight up the hill," advised Cernan. "I don't mind pitch, but I sure don't like roll."

"I don't either," added Schmitt sardonically.

Cernan updated Bob Parker in Houston with their progress. "Let me tell you, Bob, I've got to go cross-slope some of the time because the Rover is really working to go uphill now."

ABOVE LRV-1's three expeditions across the plain at Hadley took Scott and Irwin a total of 27.9km. Two journeys skirted Hadley Rille and the longest took them onto the lower slopes of Mount Hadley Delta. *(NASA/David Woods)*

RIGHT On the lower slopes of Mount Hadley Delta, David Scott was intrigued at the depth of their boot prints compared to the shallow Rover tracks. As Jim Irwin walked ahead of him, he took this picture to capture the difference. *(David Scott/NASA)*

Hillside stability

Once the astronauts had taken their vehicles onto the lower slopes of nearby mountains, they were presented with new challenges. In general, on the flatter areas of the Moon, the loose top surface of dust is quite thin, a few centimetres at most, with the dust below compacted hard by millions of years' settlement. On the lower slopes of mountains, there was a very slow but constant accumulation as vibrations from each new asteroid or meteorite impact from space disturbed the soil particles, which then tended to move downhill. As a result, the loose dust layer on these slopes was much thicker.

"Look at the Rover tracks," called Scott as he began his geology work on the lower slopes of Mount Hadley Delta. He was struck at the difference between the depth of their boot prints and the Rover's wheel impressions. "I'm going to take some pictures of the Rover tracks here. And our boot prints, both. Look at the difference. That old Rover is light."

"It does a lot better than we do!" agreed Irwin.

The design of the Rover's wheels, especially the way they flattened against the surface, meant that they were great at spreading the Rover's weight compared to the astronauts' boots. Consequently, on very soft soil, the astronauts tended to sink in much more.

As Scott took LRV-1 along the contours of Mount Hadley Delta on the second day of their mission in order to get to a boulder with a hint of green that had caught Irwin's eye, he experienced a significant amount of sideslip in the soft soil, which he likened to driving on a slope of very powdery snow. "The problem here is the slope and the softness, the lack of cohesiveness of the material," recalled Scott years later. "It's very fine grained and very loose material. And that's what makes it really hard to move around. And that's why the Rover slides." In fact, in these cross-slope drives, one end of the Rover tended to slip downhill a little more than the other, resulting in each pair of wheels making its own set of tracks in the soft soil.

When they reached the boulder and got off, they discovered that the parked vehicle not only had a tendency to slide downhill, it was sitting with an up-slope wheel off the ground. As Scott worked at the boulder to acquire a sample, Irwin held on to the Rover to keep it in place.

John Young took LRV-2 to the highest point of any Rover when they drove 170 metres up the lower slopes of Stone Mountain to the Cincos Craters on their second outing. When he started driving cross-slope, Charlie Duke discovered that being in an open car with a sporty ride could be an interesting experience for the passenger, especially if he was on the downhill side. "Okay, we're going cross-slope, Tony," called Charlie Duke. "And I feel like I'm about to fall out."

BELOW On cross-slope drives, Scott noticed how LRV-1 tended to sideslip a little due to one pair of wheels running slightly down-slope of the other. Image compiled from photographs taken by Jim Irwin. *(David Scott/NASA)*

ABOVE Jim Irwin hangs on to LRV-1, its rear wheel up off the ground, while David Scott samples a boulder that had attracted his attention on Mount Hadley Delta. *(David Scott/NASA)*

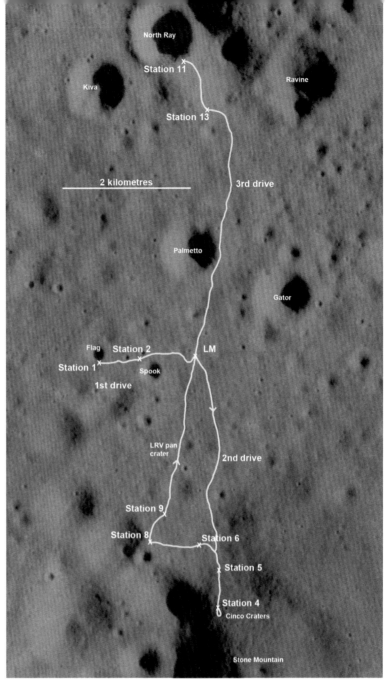

"Glad you got a seatbelt on, Charlie," said Young reassuringly.

Duke knew this was just a feeling. He recalled, "You're looking over to the side and there's no visible support there! There's no door or anything. And, so, you feel like you're going right off the side. Up or down[hill] felt real secure; but the cross-slope, fortunately, it'd dig into the dust and it'd stop; so we never felt like it was going to turn over."

Going downhill

What goes up, must come down. On the way back down Stone Mountain, Young, normally a reticent man of few words, couldn't suppress his unalloyed joy when he brought LRV-2 back to the Cayley Plain. "Ya-ho-ho-ho-ho. Look at this baby. I'm really getting confidence in it now. It's really humming like a kitten."

"Oh, this machine is super," chimed in Duke.

"Probably a good idea you couldn't see how steep it was going up," suggested Capcom Tony England, listening in from Houston.

"Darn right it was," agreed Young. "Okay. I've got the power off, and we're making 10 kilometres an hour. Just falling down our own tracks." They were running slightly sideways to turn a 12° slope into 9°, but Young had to keep on top of the steering. "Uh-oh!" he said when he tried to avoid an obstacle only to find the vehicle's momentum was fighting his attempts at steering.

"Almost spun out," said Duke.

After the mission, Young explained in his inimitable style, what going downhill in the Rover was like. "If you let that rascal loose, she'd go down that hill in a big hurry. When you got the Rover up to about 10 clicks going down a hill, it's just like riding a sled on ice. No matter which way you turn the wheel, the thing's going straight. I mean, it'd be sideways, but still going in a straight line downhill. Lot of mass there."

ABOVE On Apollo 16, LRV-2 took John Young and Charlie Duke west to Flag Crater, south to Stone Mountain and north to a relatively young crater called North Ray. Their total distance travelled was 26.6km. *(NASA/David Woods)*

Speed demons

In a straight line, the Rover typically moved at 8 to 12kmh, and Scott usually found that the back end of LRV-1 was prone to sliding sideways at speeds faster than about 12kmh. If he ever did fully open up the 'throttle', he didn't let on. On the last two missions, however, a little of the test-pilot/fighter-pilot competitive mentality came to the fore as first Young and then Cernan tried to claim bragging rights as the fastest drivers on the Moon.

"Tony, this is at least a 15° slope we're going down." Duke was narrating during Apollo 16's third drive. "Man, are we accelerating," he laughed. Their landing site in the lunar highlands was proving to be a very undulating landscape "Whoever said this was the Cayley *Plain*?"

"We've just set a new world's speed record, Houston," called Young. "17 kilometres an hour on the Moon."

"Well, let's not set any more," advised England.

What probably helped the Rover's stability at speed during the later missions was that the rear chassis was more heavily loaded with science experiments and they were also carrying a slightly greater load of rocks back to the LM.

On the return leg of the Rover's longest sortie, during Apollo 17 at Taurus-Littrow, Cernan asked for everything the Rover had as he and Schmitt descended the scarp. "What was it, 17½ or 18 clicks we hit coming down the Scarp, Jack?" he nonchalantly claimed, fully aware that Young had laid claim to only 17 clicks. Schmitt just laughed at the suggestion.

In fact, Cernan believed that even with a more capable Rover, it would have been hard to exceed the speeds that the Apollo astronauts achieved. "I don't know that it would have been advantageous to have had a Rover that could have gone any faster," he reflected in the years after his mission. "You needed to be careful in driving and probably couldn't have used the additional capability. When we set a speed record coming down a hill, we were really hauling the metal; we were really tearing. And, not being able to see all that well ahead, I don't think we would have wanted to come down much faster.

"Of course, we were travelling over uncharted terrain," he continued. "And, if we'd been on a surveyed path with no boulders or potholes in the way – then we could have driven comfortably at a lot higher speed. But if we'd hit a boulder at 10 kilometres per hour and smashed the hell out of the front end of the Rover, we'd have been out of business. We could have gotten back to the LM, but we wouldn't have gotten the job done that we'd gone to the Moon to do."

Suspension and ride

Wherever the astronauts took their Rovers, they came across a surface that was constantly undulating at all scales, thanks to the way meteoroids had blasted craters large and small into the regolith. A fresh crater would be sharp and clear, and usually surrounded by blocks that were formed instantaneously by the enormous energy of the impact – so-called 'instant rock' or 'breccia'. Over time, subsequent impacts and the constant rain of micrometeoroids would erode the blocks back to dust and smooth off the crater's profile while adding more craters on top. This made for a surface that was rarely flat and always a challenge for the driver and the Rover's suspension.

Like a racing car that had lost its rear wing, the Rover lacked downward force, entirely

BELOW The travels of LRV-3 ranged from one side of the Taurus-Littrow valley to the other for a total of 34.8km. The second day's huge drive of over 20km took Gene Cernan and Jack Schmitt 7.5km from the Lunar Module. *(NASA/David Woods)*

through the weak lunar gravity. So when an upward bump in the ground lifted the vehicle, the momentum of the ¾-tonne fully loaded vehicle kept it going, often lifting one or more of the wheels off the ground to make a short 'airborne' coast before coming down again.

"Man, this is really a rocking-rolling ride, isn't it?" laughed Scott on his first outing with LRV-1.

"Never been on a ride like this before," agreed Irwin.

"Boy, oh, boy! I'm glad they've got this great suspension system on this thing. Boy." Scott found the ride to be a lot more bouncy than he had expected from his drives in the 1G Rover that they used for training. In his post-flight debrief he said that, "the 'floating' of the crewmembers in the sixth-g field was quite noticeable in comparison to 1-g simulations." He continued, "At one point during the 'Lunar Grand Prix', all four wheels were off the ground, although this was undetectable from the driver's seat."

The Apollo 15 Lunar Grand Prix was intended as a filmed test of the Rover for the benefit of engineers. Scott would put the car through its paces in front of Irwin's 16mm cine camera so that the turning of the wheels and the movement of the suspension as well as the whole vehicle could be analysed on Earth. Unfortunately, Irwin's camera jammed and so he became the only spectator at Scott's Grand Prix.

Eventually the engineers got what they wanted, thanks to Charlie Duke's filming of John Young taking LRV-2 for a spin. They could finally make their measurements to test their predictions of wheel slippage and suspension travel. That film has become an iconic favourite of documentary makers in the years since, because it is the only 'spectator' footage of a Rover being driven in its intended environment. Readily visible are the huge rooster tails of dust that Pavlics's wire wheels could throw up and the floating ride it gave its passengers as the vehicle bounced off the surface.

The Grand Prix film wasn't properly representative because it had only one astronaut aboard. But even with two passengers weighing it down it was still quite bouncy, as Duke found out when he and Young approached the lower slopes of Stone Mountain. "Hey, that was super," he yelled. "That wheel just left the ground."

"This is the wildest ride I was ever on," laughed Young.

"I love it. It's great!" cried Duke.

Years later, he recalled the ride. "It was like you were in slow motion as it bounced, because of the one-sixth gravity. We didn't spend a lot of time in the air, but you had that springy feeling that was more a slow motion, slower frequency because of the light gravity."

"Man, this is a fun ride," continued Duke on the Moon. "Occasionally, the back end breaks loose, but there's no problem. This is really some machine."

"It's just like driving on snow, Houston. By golly!" added Young.

"Gee, I know all about that," opined Tony England in Houston. England, a scientist, had done research in Alaska and Antarctica in the use of radar to probe the depths of ice. He was used to driving over snowfields.

"I know you do," chuckled Young, who had adopted the very unsnowy Florida as his home state. "But us Florida boys don't know much about it."

Cernan made comparisons with Earthbound cars when he described the bounciness of the Rover's ride. "You often couldn't see small craters and they were important in the same way that potholes are to you when you're driving down the highway. The difference is that, in one-sixth gravity, when you do hit a boulder or you do drop one wheel in a small crater, you literally lift the Rover off the ground. You were literally driving on three wheels a good part of the time. It wasn't a choppy, sports-car feel,

ABOVE This still frame from Charlie Duke's Grand Prix film shows LRV-2 with at least one wheel off the ground as John Young puts it through its paces. The huge rooster tails of dust thrown up by the wheels are readily apparent. *(NASA)*

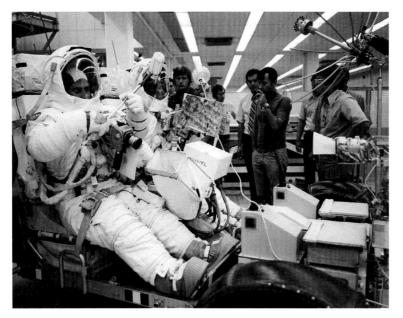

ABOVE Four months prior to flight, Apollo 17 astronauts Jack Schmitt and Gene Cernan are on the 1G Rover during training. John Young is talking to them. Schmitt is wearing his seat belt, which clips onto the outboard handhold. *(NASA)*

driving down the highway going bang, bang, bang. It was more of a waltz. Boom, and you go u-u-up and come down. Hit another one; u-u-up and come down. And, of course, not only did the Rover bounce but so did we. So it was a very good thing that we had snug seat belts. Without them, we could easily have bounced out of the seats."

All the Rover astronauts felt that the seat belts were vital to keeping them on board. "It really cinched you in there," recalled Duke. "It had a good over-centre position and it was easy to operate. It had a big hook on it so you could just reach over and hook on the outside handle of your seat, which was real easy to do. You really couldn't see it [from inside the helmet]; you were working blind. And, then, that big handle would give an over-centre lock and it really cinched you in there firm."

David Scott used his seat belt for some geology. As they drove home to conclude their first day's excursion, Irwin was busy describing apparent layers on Mount Hadley to their north. "There's a definitely linear pattern that looks like it dips 30° to the west." Just then, the Rover came to a halt. "How come we stopped?"

"I got to put my seat belt on," explained Scott. Somehow it had become unhooked. Mission Control were carefully timing their journey to try and understand the Rover's performance on this, its first outing, and they were eager to know about any stops and starts.

"Dave, stand by for mark when you start," called Allen. "Help us on our speed calculations."

"Yes, I'm sorry about that, Joe," said Scott, "but I'm pretty unstable without that seat belt and there's a lot of feedback into the controller." The side-to-side movements of the Rover as it pitched and rolled on the uneven surface could cause an astronaut to make unintentional steering commands if he wasn't tightly held in place. Scott got moving again and let Allen know.

"Anytime we stop, Joe, I'll let you know," laughed Irwin.

A few metres on, Scott caught sight of an appealing rock. "Oh, there's some vesicular basalt right there, boy. Oh, man," he exclaimed. Scott had taken geology to his heart, and wanted to collect the rock but felt that Mission Control would not allow them the time to stop and sample it properly.

"Hey, how about..." Scott stopped mid-sentence as he realised where this was going. Quickly deciding to use the excuse of a second seat belt adjustment as a reason to stop, he pulled up close to the rock.

"Okay; we're stopping," said Irwin, keeping his promise to Allen to let them know of stops and starts.

"Let me get my seat belt," fibbed Scott. "It keeps coming off."

Without letting on, he undid his seat belt, got off the Rover and gathered his tongs to grab the rock; making sure to properly document the process with his camera. The rock, now known as the 'Seat Belt Basalt', is a type that once contained lots of dissolved gas that created the many holes called 'vesicles' which had caught his eye.

This episode would lead to the development of a long-handled tool to allow subsequent astronauts to quickly pick up so-called 'opportunity samples' on future missions without removing their seat belts or having to get off the Rover.

The Rover's bounciness, coupled with the very bumpy terrain, meant that its drivers had to take into account one other factor: the communications unit, the LCRU, which projected out from the front of the vehicle. It had to be detachable, yet when mounted on the Rover it had to present a set of space radiators to space. There had been nowhere left to mount it but hanging off the front of the forward chassis. Consequently when driving into a small crater, it was wise to take things slowly for fear of the suspension bottoming out and causing the LCRU to slam into the opposite upslope.

The T-handle

Once on the Moon, the commander always drove the Rover. Eric Jones, noted historian of Apollo's exploration of the Moon and editor of the definitive *Apollo Lunar Surface Journal*, put it like this: "Naturally, none of the three J-mission commanders ever relinquished control – nor did any of the LMPs breach etiquette by asking to drive."

Whether or not the commander's bragging rights were the issue, it was David Scott who was in control of LRV-1. In his first seat-belt stop, he had noticed a problem with the design of the Rover's T-handle; it was prone to being unintentionally deflected, either by the driver being swayed from side to side or by being knocked by the passenger.

Accidental deflection aside, the T-handle had functioned well on its first outing. With the relaxed arm position to drive it, Scott had benefited from resting his right forearm between stops in readiness for the next bout of geology.

He commented on the T-handle on his way back to the LM at the end of their first outing: "It's easy to drive; no problem at all. Very responsive. I can put the throttle right up to the stop or [indeed] at some intermediate position; and take my hand off and rest my hand. If I want to go left or right, I just put a little pressure until I get the angle I want and then let it off and we re-centre on the steering. It's really neat." His use of the term 'throttle' reflected Earthbound internal combustion cars of the era. When a driver wanted to go faster, his accelerator pedal operated a cable that literally opened a throttle valve that allowed a greater flow of air into the engine. As the air went through the carburettor, it picked up more fuel, increasing the power of the combustion.

Overall, the results of the T-handle's first drive were excellent and the hard work by Marshall and their contractors to replace the original joystick so late in the day had paid dividends (see Chapter 5). But the accidental

ABOVE Schmitt and Cernan in training on the 1G Rover four months before their flight. Schmitt, nearest the camera, is using a sampling tool to practise taking opportunity samples from on board the Rover. *(NASA)*

deflection problems of the slightly cramped crew station would not be easily fixed, and when the issue emerged again the following year Apollo 16 commander John Young had his own fix. As Young and Duke were driving to their first geology stop, Duke was busy dealing with maps, taking photographs and generally moving about in his seat.

"Quit hitting my arm!" complained Young as Duke's movements caused an unwanted input to the steering.

Duke recalled the situation: "As I moved around in my seat, I'd keep hitting John's arm and it would knock him over. And when I hit his arm, it would move the handle and it would turn the vehicle. And he was getting a little frustrated with me, for which I don't blame him. But it's really hard to stay out of his way. You had plenty of room as far as fore and aft goes; but, side-by-side, we were sort of shoulder-to-shoulder and it was difficult to stay out of his way."

Blinded by the light

Imagine one morning driving off-road in an open car with a low Sun shining brilliantly in the sky, and you have to travel east, staring straight into the Sun. It would normally be an unpleasant experience as you squint into the Sun to see where you are going. Now imagine travelling west with the Sun behind you. Under normal conditions on Earth, you would expect the drive to be much easier. But astronauts driving on the Moon found this to be a very difficult task.

"Driving down-Sun in zero-phase is murder," commented Young as he struggled to pick out a path among the craters.

"It is, isn't it?" sympathised Duke.

"It's really bad," said Young.

Zero-phase referred to the direction that is 180° from the Sun, i.e. with the Sun right behind you. And on the Moon it proved to be astoundingly bright. "Zero-phase just basically washes everything out," recalled Duke, "the craters, the slope, and, also, the smaller rocks. And then they would start coming [at you] through the glare."

Young had much to say about it in the post-flight debrief. "Man, I'll tell you, that is really grim. I was scared to go more than 4 or 5 kilometres an hour. Going out there, looking dead ahead, I couldn't see the craters. I could see the blocks [of rock] alright and avoid them. But I couldn't see craters. I was scared to go more than 4 or 5 clicks. Maybe sometimes I got up to 6 or 7, but [when I did so] I ran through a couple of craters because I flat missed them until I was on top of them."

Surprisingly, Young found the return journey much easier, even though he was then driving into the Sun. With his gold visor down the raw intensity of the Sun, whilst still not ideal, was no worse than driving sunwards on Earth wearing shades.

"The other direction was about twice as good. I saw my tracks on the way back. We were doing 7, 8, 9 and 10 clicks."

Three effects worked together to make driving down-Sun difficult, and they can also be can be seen on Earth when the Moon is full. People have long noticed how the Moon becomes much brighter close to its full phase. This is not only because the Earth-facing side is fully illuminated, with the Sun behind us, but also because all the shadows in the lunar landscape have disappeared. And without shadows to darken what is being seen, the landscape appears brighter when viewed from Earth. On the Moon, when objects in the direction directly down-Sun hide their own shadows, it is difficult to sense the shapes of the craters and rocks in the washed-out scene of monotonous grey.

A second minor effect is due to the fact that

some of the soil particles are tiny glass spheres which act as retro-reflectors – much like road signs that brighten in headlights as they return light back in the direction from which it came – further brightening the scene.

A third important effect that doubles the intensity of zero-phase light is a phenomenon called coherent backscatter which was only understood more than a decade after the Apollo missions. When light passes through a medium consisting of countless reflecting bodies, it is scattered, just as we would expect. The powdery nature of lunar soil would be a good example of such a scattering medium. However, as the light bounces around within the material, quantum effects between the light rays cause them to reinforce each other coherently and this is strongest in the direction from which the original light beam arrived. Whether on the surface or in lunar orbit, astronauts noted how bright the Moon's surface appeared directly opposite the Sun and this brightening is also very apparent here on Earth when the Moon is full.

Breakdown and recovery

There was one occasion when all the bouncing about did appear to lead to a breakdown of the Rover; on this occasion Young and Duke were 3km from the LM. They were making their way around the side of Stone Mountain to a planned geology stop they called Station 8 and they decided to climb a steep ridge, but found that progress was slow.

ABOVE This panorama, stitched from images taken by Jack Schmitt, shows where he and Cernan had set up Apollo 17's science experiments. It gives an impression of how the landscape brightens and becomes indistinct near the zero phase point, within Schmitt's shadow. *(Harrison Schmitt/NASA/David Woods)*

"Have you got full throttle on?" asked Duke.

"I got full throttle," replied Young.

"Boy, we're hardly moving."

Young glanced at the ammeter and saw what the problem seemed to be. "We've lost the rear-wheel drive. Not reading any amps on the rear wheel."

"John, why don't we check it?" suggested Duke. "It just might be a steep slope. But the front wheels were really digging in."

"No, Charlie. The ammeter was reading zero!" returned Young. If the slow-down had been due to the slope, they would expect the meter to show a large current.

"I know. Could be a broken meter," added Duke.

Tony England suggested they avoid climbing to take the strain off the front motors. Two minutes later, they came upon a collection of boulders and decided to make these Station 8. Boulders that might have been ejected from a large nearby crater were what they were after anyway, and this stop would give them a chance to diagnose why no power was getting to the rear wheels.

Young had an initial theory. "My best guess of what may have happened, Houston, is that we cut a wire or something on the back."

"Cut a wire?" questioned England. This sounded serious.

"Yeah, a wire going back there to that aft thing." Young was aware that their drive had been fairly rough on the vehicle. "On the way down here, when we were bouncing up in the air, we came down on at least two rocks that I know about." Young was likely sensitive to the idea of cut wires because on their first day, he had accidentally got his feet caught up on cables that led to a science instrument and had irreparably yanked them out. He still felt bad about wrecking the experiment.

Unfortunately it was difficult to check the Rover's wiring harnesses. They were hidden by the underside panels and thus well protected, so it was unlikely a rock would have penetrated them to sever a wire. While the astronauts set about a period of field geology at the boulders, the LRV engineers in Houston considered the problem.

Twenty minutes or so later, Young jumped back into his seat on the left and began troubleshooting. First he made sure that both steering systems were working. They were. Next he checked that each motor was being fed from the correct power bus. Battery 1 fed busses A and B, battery 2 fed C and D. The front wheels could take their power either from bus A or C while the rear wheels were powered from bus B or D. Front-wheel power was no problem but the rear wheels failed to respond regardless of whether they were switched to receive power from B or D. That ruled out it being an individual bus or battery problem.

If it wasn't the busses or the batteries, then perhaps it was the devices that controlled the power to the wheels. The wheel motors did not receive their power directly from the battery, but via a switch which was controlled by two so-called 'pulse width modulators' (PWM1 and PWM2) to allow the electric motors to be

ABOVE John Young at Station 8 on the side of Stone Mountain. At this point, they know that LRV-2 has a problem but they will complete their geology before addressing it. As Charlie Duke takes the photographs that make up this panorama, Young is aligning the high gain antenna to let Houston use the TV camera. *(Charlie Duke/NASA/David Woods)*

RIGHT Schematic diagram of the Rover's system for delivering controlled power to the wheels. LRV-2's problem lay in the fact that Duke had inadvertently knocked the 'PWM Select' switch to '1' rather than 'Both'. *(David Woods)*

efficiently 'throttled' up and down (see Chapter 3).

The throttling of each of the four motors could be controlled by either PWM, dictated by a set of four switches on the lower right of the Control and Display panel called 'Drive Enables'. The normal setting was for the front wheels to be controlled by PWM1 and the rear to be controlled by PWM2. Each wheel motor could also be disabled by setting these PWM switches to neither 1 nor 2.

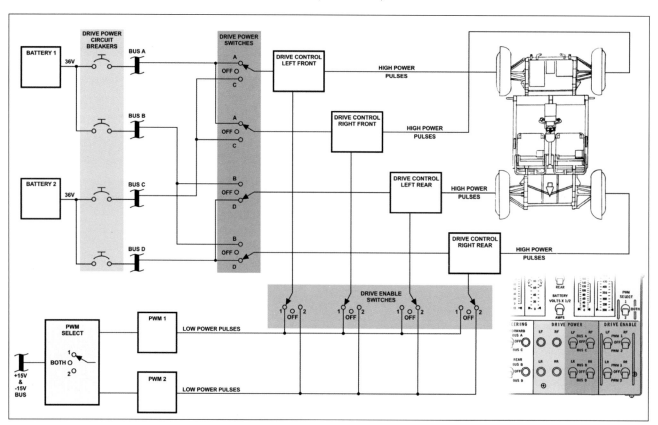

"Okay, John," called England as the troubleshooting progressed. "We'd like you to try the PWM1 on the Left Rear and Right Rear."

"Okay, I'm in PWM1 on Left Rear and Right Rear?" confirmed Young, glancing at the four 'Drive Enable' switches.

"Right," agreed England, going on to clarify Mission Control's next step. "We would like all the Drive Enables to PWM1."

Directly above the Drive Enable switches was another switch called 'PWM Select', which controlled the power supplies for the PWMs. It had three positions that allowed the astronaut to switch on either PWM1, PWM2, or both. "Do I have to be in 'PWM Select 1', or 'Both'?" Young enquired with his mind on this switch. He immediately saw why the rear wheels were not working. With the switch set to PWM1, PWM2 was not receiving power and so neither were the rear wheels.

"Oh, that's the problem!"

"That's the problem, you weren't in 'Both', huh?" clarified Duke.

Young was elated. "That is the problem! Somehow this guarded switch got moved. Oh, isn't that amazing."

Most likely, as Duke had tried to handle the maps, cameras and the low gain antenna, he had inadvertently hit the switch, despite its guard-rails. Little wonder, though. The terrain was rough and bumpy and the little car and its passengers were thoroughly tossed about in the light gravity.

This wasn't quite the end of the affair. In troubleshooting the problem above, it seems that for quite a while after Station 8 the Rover was driven without power applied to the front wheels. Without power to two of the wheels the navigation system could no longer work out which was the third fastest wheel, since only two wheels were delivering odometer pulses. Until the system was reset at Station 10, Young and Duke had no valid navigation information. In this case, it wasn't a problem. Though they were 3km away, they had reached all their planned sites for the day and, being on high ground, they could see across the plain to their Lunar Module.

During the three Apollo Rover missions there were two other failures of LRV equipment that plagued the vehicle; one minor, one major. The

LEFT This detail from a photo of David Scott on LRV-1 shows the tool pallet at the back. The small vice mounted at the top is arrowed.
(Jim Irwin/NASA)

minor problem occurred towards the end of Apollo 15's time on the Moon. David Scott had struggled with a large cordless coring drill with which he had drilled holes in the lunar surface to insert thermometers that would take the Moon's temperature. For the final hole, he used a hollow drill stem to gain a core of soil nearly 3m deep that would be a record of Hadley going back about a billion years.

Extracting the long drill stems was far more difficult than expected, and both Scott and Irwin had to get their shoulders under the drill and heave with all their might to make the core budge. Scott even managed to sprain his shoulder, which is a remarkable testament to the trust that they placed in their suits.

The drill stem came in six sections, each 50cm long. To take them apart, Scott used a self-gripping wrench like a plumber would use to turn one section while the other side of the join was held in the jaws of a similar wrench installed like a vice on the pallet at the back of the Rover which carried the hardware they needed for their geology. Every time Scott tried to turn the stem, the jaws on the vice would refuse to hold it.

"I hate to tell you, Jim, but that... Oh boy!" Scott was getting exasperated. The drill had already caused much time to be lost and he could see that their final drive was about to be curtailed. "This vice is on... I swear it's on backwards."

On their first day, Irwin had installed the vice onto the Rover's pallet but it was designed so that it could only be installed one way and therefore he couldn't have made a mistake.

ABOVE Jack Schmitt took the images for this panorama at the start of Apollo 17's final day on the Moon. Gene Cernan is preparing LRV-3 for their drive to the North Massif. *(Harrison Schmitt/NASA/David Woods)*

BELOW Charlie Duke and John Young during training with the 1G Rover in January 1972. Young, with the red stripes, has a geology hammer in his shin pocket that is very near the Rover's rear fender. This photo shows how easy it was for the commanders to inadvertently catch the fender, given the limited downward visibility they had past their chest-mounted control unit. *(NASA)*

It turned out that the original drawings for the pallet had been in error. When the 1G Rover was set up, someone spotted the error and fixed it but the change never made it to the Lunar Rover. Unable to loosen the drill stems with the backwards vice, Scott and Irwin grabbed the stem by hand, with its sharp helical flutes which could have punctured their pressure gloves, and managed to undo three of its sections. But no matter how they tried, they could not separate the remaining sections. It was decided to bring them home with their precious soil contents as one 1.5m long section; somehow finding space in the cramped spaceships to stow it.

The only Rover failure that came anywhere near to causing serious concern was when the vehicle's fibreglass fenders were inadvertently damaged. Each wheel's main fender was bolted to the motor drive's casing but there was a deployable smaller section which was attached to the main fender on rails that allowed it to be slid into position after the wheels were deployed. These extensions were vulnerable because they were lightweight and thin and because they were at the very corners of the Rover where the two astronauts regularly busied themselves in their bulky stiff suits with limited visibility and feel.

A fender extension was lost on all three missions, though in the case of Apollo 15 no one noticed it until the pictures were returned. It never became a nuisance because it came off the front wheel and any dirt being held in by it was already travelling down and forward, away from the Rover. It was the rear fender extensions that were a problem because the dust they controlled was otherwise launched up and forward over the Rover and crew.

LRV-2 lost its right-rear extension towards the end of the second day when Young and Duke were swapping sample bags at the end of a rock collecting exercise. As Young went to move around the corner of the Rover, his leg caught the extension. "There goes the fender," he noted.

"Uh-oh," said Duke.

From that point on, the two astronauts found that they had to brush down themselves and the Rover at every halt. They knew that in training, it was easy to damage these extensions, but hadn't appreciated how important they would be on the Moon. Dust is a problem because it fouls up the mechanical areas of the suit; the neck ring and the rotating glove joint in particular. Being dark in colour, it also affects the ability of the radiators to dissipate heat by allowing the Sun to warm the very surfaces that ought to be cooling. Apollo 16's problem with the broken fender presaged the problems Apollo 17 would have when they lost one of theirs – and this time it happened at the worst possible moment.

Cernan and Schmitt hadn't even begun their first foray away from the LM. They had LRV-3 deployed and Cernan was fussing around getting it loaded up and operating a gravity experiment mounted on the rear. As he shifted around the right-rear corner of the vehicle, a hammer in his shin pocket caught in the fender extension and, before he realised what was happening, the fibreglass extension was torn off its rails.

"Oh, you won't believe it," moaned Cernan as he surveyed the damage. "There goes a fender."

"Oh, shoot!" said Schmitt.

Cernan took some time to finish the task that he was doing and weigh up his options. "I hate to say it, but I'm going to have to take some time to try to get that fender back on."

Bob Parker in Houston queried further. "Was it the rear fender, Geno?"

"Yeah. Caught it with my hammer, and it just popped right off." Cernan was aware that even with the fender, a little time had to be spent dusting the radiators throughout the day. Now he contemplated the extra precious time that would be lost trying to keep the tenacious stuff at bay. "I can't say I'm very adept at putting fenders back on. But I sure don't want to start without it."

Thankful that the NASA team had had the foresight to ensure that a roll of grey duct tape was handy, Cernan fruitlessly set about trying to effect a repair. "Good old-fashioned American grey tape doesn't stick to lunar-dust-covered fenders. One more try."

The dust was defeating him as both parts of the fender were already covered. It was hard enough to get the tape off the roll with his gloves, never mind trying to get it to stick to the dusty fender. Cernan spent ten minutes trying to repair the broken fender before declaring to Houston an optimistic victory. "Bob, I am done! If that fender stays on, I'm going to take a picture of it because I'd like some sort of mending award. It's not too neat, but tape and lunar dust just don't hang in there together."

Cernan and Schmitt set off on their first drive, which was to a crater about 1km away. The fender repair seemed to be going well but as they returned to the LM later, Schmitt realised that dust was falling all around them. "I think you've lost a fender. I keep getting rained on here."

"Oh, no!" Cernan glanced to his left to see the Rover's shadow and spotted the evidence of a large rooster tail of dust coming off the right-rear wheel. "It probably didn't stay. I can see it in the shadow."

"Sure, look at it," said Schmitt.

"Oh boy, that's going to be terrible." Cernan thought about the effects he knew the dust would have on the Rover's thermal systems, the suits, their cameras with their adjustments and even their helmets and visors which could easily be scratched by attempts to wipe the dust off. "Man, I hate this dust. I got to make a new fender tonight."

Years later, Cernan recalled how important it was to repair the fender. "It was just an unacceptable situation. Imagine, a little thing like knocking a Rover fender off having the potential of compromising the rest of our mission. I still have no doubt that we had to find a fix. If we

RIGHT A solution to LRV-3's broken fender. This was the result of some overnight thinking and testing. Left to right are Apollo 16 astronauts John Young and Charlie Duke, their boss Deke Slayton, Apollo Program Director Rocco Petrone and engineer Ronald Blevins. The instructions for building this repair would be read up to the crew before the start of their second day. *(NASA)*

BELOW This incidental photo by Jack Schmitt shows the freshly installed fender repair just before they set off on their longest drive to Station 2. Gene Cernan is already seated and is initialising the navigation system. *(Harrison Schmitt/NASA)*

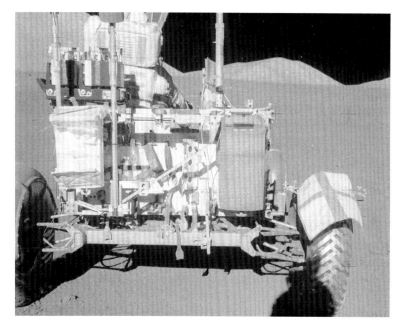

hadn't, then at bare minimum it probably would have taken 30 to 40 per cent of our time to consciously keep the dust off of the Rover, the tools, and the rest of the equipment. We wouldn't have been able to drive as fast. We would have had to spend a lot more time dusting."

By the time the two astronauts had got back into the LM and out of their suits, Terry Neal, a crew systems engineer who was very aware of all the bits and bobs that were at the astronauts' disposal, and Ron Blevins, a training instructor, were coming up with a plan for a repair. Joe Allen explained it to the crew: "We've been doing some thinking down here about how to fix the fender. And it's going to involve using utility clamps, from inside your LM there, instead of tape, to fasten some sort of stiff material onto the Rover in place of the missing fender."

While Cernan and Schmitt rested on the Moon, John Young donned a suit in Houston to test the plan, and then after the crew had woken he read it to them. The procedure for a Rover fender repair on the Moon is as follows:

- Take four "cronopaque" pages from the lunar surface maps book. These stiff pages are about 20 by 25cm.
- Use grey duct tape to join these pages together to create a single sheet about 38 by 27cm. Take care to exclude air bubbles that would expand in the lunar vacuum and lift the tape.
- Take the taped card out to the surface along with two utility clamps from the LM cabin.
- Lay the taped card over the main structure of the remaining fender so that it sits rearward from the fender by at least 10cm.
- Use the two utility clamps to fasten the replacement fender to the remaining fibreglass fender so that it naturally forms a 'half-pipe' shape. The guide rails for the fender extension will be a good point to clamp onto.

The utility clamps were in the LM cabin to provide mounts for little lamps that the astronauts could attach to bars that protected a centrally mounted telescope.

The fender repair proved to be reasonably successful and lasted right through the Rover's longest drive. It was only at the furthest reach of their final drive that it showed signs of wearing out.

"Our fender's beginning to fade," noted Cernan at Station 8, a spot at the base of some nearby hills 4km northeast of the LM. "And, uh-oh, the clip came off." The clamp holding the card fender on the inside of the wheel had come adrift. "We'll have to fix that before we start. The outside one's holding, but the inside one's not."

By the time they reached Station 9, the last on their journey, and the last in the Apollo programme, they had begun to notice the rain of dust had started. By now, the fender

RIGHT Schmitt is seated on LRV-3 and they are at Station 2, as far from the Lunar Module as a Rover ever got. Cernan took this picture to show how his handiwork had stood up to the 9km drive. *(Gene Cernan/NASA)*

replacement wasn't just flapping about, it was being pulled under the original fender by coming into contact with the wheel.

"Boy, everything is really bad now," said Schmitt.

"Yeah, the fender dug under," noted his commander. "See if you can straighten it out."

The state of the dust-covered Rover was what Cernan had been trying to avoid throughout their travels, and for the most part, he was successful. "That's just a sample of the kind of dust we would have got, Jack, if we hadn't have had that fender yesterday. Fender's almost worn out."

Before Cernan left the surface for the last time, he undid the clamps and rescued the fender to take back to Earth. He then asked Capcom Bob Parker, "Hey, congratulate José on that fender, will you?" José was a nickname for John Young. "Because I think he just saved us an awful lot of problems. He and whoever else worked on it."

"He [John] mumbled something very humbly about a thousand guys," informed Parker who also understood that this repair had been another team effort.

"Well, tell him that's going to be my 'bring home' present to him."

Cernan's fender fix is now on display in Washington DC's National Air and Space Museum.

Keeping cool

There were two thermal issues that had concerned the engineers in the months leading up to Apollo 15; the temperature of the motors, which proved not to be a problem, and the temperature of the batteries. Their concern was expressed in their provision of a warning flag that was mounted at the top of the instrument panel. Normally, this spring-loaded hinged plate stayed folded against the casing, held in place by a magnet. But if sensors in

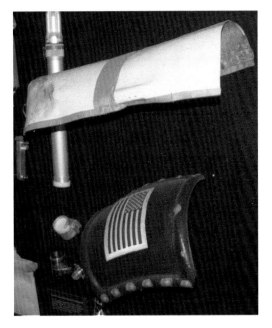

ABOVE LRV-3 is at Station 5 at Camelot Crater near the end of a 20km drive. Schmitt's photo shows how the replacement fender is starting to curl at the tip as it succumbs to the conditions. *(Harrison Schmitt/NASA)*

LEFT Cernan's replacement fender on display with one of the fender extensions at the National Air and Space Museum. *(David Woods)*

the motors or the batteries indicated excessive temperatures, then a pulse was sent to an electromagnet which released the flag, causing it to pop up in front of the astronauts.

In general, battery 2 was warmer, even at the start of the Rover's mission, primarily due to not having a wax tank attached as on Battery 1.

On Apollo 15, for example, the batteries started out at 20° and 26°C respectively, and by the end of the last day they were 42° and 45°C. The dust covers were left open while the astronauts slept with the expectation that they would have closed automatically by the start of the next day. Only once did this happen, with battery 1 on the first crew rest period. Over the second rest period there was almost no cooling. Even a small accumulation of dust from day 2 and the inability to clean the radiators kept them from working, but thankfully the temperatures stayed within limits during the final day's driving.

The dust problems on both Apollo 16 and 17 meant that their batteries began to warm excessively after they lost their fenders. Added to this was the fact that it had also been decided to run the Rover's communications system from the LRVs' batteries and the greater discharge this demanded increased the heat generated by the batteries themselves.

On Apollo 16 there was very little overnight cool-down, due to small amounts of dust on the radiators and the fact that the Rover had been parked too near the Lunar Module, reducing their exposure to space. Mission Control warned Young towards the end of his last drive to expect a temperature warning for the batteries, and when it came he merely reset it even though it reached nearly 62°C.

Apollo 17's LRV batteries had endured unexpected heating during transit to the Moon. But the LRV thermal engineers knew about this and prepared to have the astronauts open the dust covers over the radiators during the first EVA.

The loss of LRV-3's fender extension on Apollo 17 early in the mission made a bad initial situation even worse because the batteries were already hotter than any previous LRV batteries (35° and 43°C) at deployment. The constant battle with dust combined with the fact that LRV-3 was a more heavily loaded machine exacerbated the problem.

"Okay, we got a flag on the Rover," said Cernan on the second day. The Sun would be even higher on the third day. "I'm reading 136[°F, 58°C] on battery number 2."

But Mission Control had already noted that the dust problems of the first day had resulted in elevated temperatures and had asked the Rover engineers to advise what temperatures could really be tolerated, and the answer was 140°F, 60°C. They were eating into their safety margins and they would have to watch the temperatures carefully. The surrounding landscape was going to reach an average temperature of nearly 70°C before the mission was over.

Unfortunately, the temperature sensor for battery 2 failed throughout the final day and its thermal performance had to be worked out based on battery 1's readings and past experience. As Cernan parked the Rover in its final parking place, Battery 1 gave one last thermal shout via the warning flag. "I got a flag on the battery, 139°[F, 59.5°C]," called Cernan. Engineers reckoned battery 2 reached 64.5°C.

Being able to rely on battery redundancy allowed mission managers to approve LRV operation at battery temperatures above their upper operational limit.

BELOW LRV-3 parked outside as seen from inside Apollo 17's LM. Cernan has deliberately parked it side-on to the Sun and not too near the LM for optimal thermal control. *(Harrison Schmitt/NASA)*

Juggling batteries

After LRV-1 had completed its mission, engineers at Boeing were happy to discover that they had over-estimated its power requirements. It had used barely 2 of an available 8kWh. This allowed the electrical power team to reappraise how they would power the Rover's communications electronics, known as the LCRU. This unit had its own batteries; one for each of the three days, to allow it to be powered while being carried back from a disabled Rover in the event of a mishap. But the LCRU could also be powered from the Rover's main batteries.

Once they understood the Rover's actual power needs, the engineers realised that they could save over 4kg on the Lunar Module by omitting one of the three LCRU batteries and instead drawing on some of the spare capacity in the Rover itself. The landing mass of the LM was critical. It took a lot to get every kilogram to the Moon and each reduced the spacecraft's ability to hover while the commander looked for a safe place to land.

Even though running the LCRU off the main batteries imposed a greater load than originally intended, the Rover's total power efficiency exceeded expectations. By Apollo 17, it had been expected that the batteries would supply a total of 3kWh over the entire mission. However, even though at 34.8km the total distance covered was much greater than for the previous missions, only a total of 2.6kWh were consumed. At this rate the Rover could have been driven at least another 70km.

The blessings of the Rover

Throughout the cumulative nine days on the lunar surface that the three Rovers were operational, the astronauts found the LRV to be a priceless piece of kit; tough, versatile and capable. Even beyond the initial expectations that had been held by those who created it, new tricks and unexpected benefits were found.

As well as extending the reach of the astronauts to a spread of scientifically interesting sites, all of which would have been utterly inaccessible to a walking crewman, the Rover allowed them to rest a while between stops, at least physically. They would arrive rejuvenated for their next bout of geology. A more important benefit was that a resting astronaut was one that was using less oxygen because his metabolic rate was lower. Therefore he produced less heat and could reduce his cooling requirements. Since the limiting consumable on the backpack was water for the cooling system, the Rover effectively increased the safety margins for the backpack.

On Apollo 16, after their first drive, a short 1km dash west to Flag Crater, Young and Duke commented on this effect. "I tell you," exclaimed Duke, "when you go to get on this thing, you better turn your cooling down or you'll freeze."

"Yeah, I should have gone to Minimum cooling," agreed Young.

Each of them could reach back with their right hands to a control on the backpack that let them set their cooling level. It altered how much water was fed to a porous plate where it formed ice that sublimated into space. This cooled another circuit of water that ran in thin pipes around the astronaut's body. To allow access to this control while seated, the seats on the Rover had been designed with a cut-out.

During their long drive out on day two of Apollo 17's mission, Cernan and Schmitt discovered a difference in their individual cooling requirements. "That Min cooling is just about right, isn't it," said Schmitt. "No, it's just about warm for me," said Cernan.

In fact, throughout their time on the Rover, it turned out that Cernan's metabolic rate was usually a little higher than Schmitt's as he was concentrating on the driving. "Bob, is my PLSS cooling working all right?" he asked.

"Rog," replied Bob Parker in Houston. The control team around him were keeping a close eye on the crew. "It looks like it's working to us."

"Being on the Rover was one of the few places where you could relax," recalled Cernan. "Physically, you were constrained and had to relax. The only thing I had to do was drive and observe as best I could. When you're talking – particularly in a case like this, you've got to think; and thinking takes energy."

Another use for the Rover was as a moving camera platform, though not with the television camera. Except for two moments, TV could not be sent from a moving Rover because the dish antenna would lose its alignment to Earth. But the astronauts carried two other types of cameras; a 16mm movie film camera and two 70mm Hasselblad stills cameras. The original intention had been for crews to film the landscape during their drives with the movie camera but this failed on Apollo 15. On the subsequent missions, the astronaut sitting on the right made a point of photographing the scene every 50m or so as they drove in order to provide geologists with a record of the landscape between stops.

Then on Apollo 16, Young and Duke came up with another neat idea. At their geology stops, once they disembarked from the Rover, it was customary for one of the astronauts to use his Hasselblad to photograph the scene as a panorama by taking a sequence of shots and turning between each so that there was an overlap and the photographs could subsequently be stitched together.

On their way back to the LM on their second outing, Young and Duke came upon a huge depression that took their interest.

"That's really going to be a steep slope, if we go straight into it," exclaimed Duke in his running commentary to Houston, "but John is adroitly manoeuvring around it."

"I'm not going down that critter," laughed Young.

"That is really steep. Look at that," said Duke. He then began to describe the feature to Houston. "It's a subdued crater without any rim at all. It is sort of oblong. Tony, you remember out in Hawaii, at Kapoho where we saw those very small little sink-hole craters? This looks like a big one of those."

"Okay. It'd sure be good if you could swing the DAC over that way, if it's still running," suggested England. The DAC was a Data Acquisition Camera; NASA-speak for the 16mm movie camera.

Duke tried to turn it around on its mount to film it but found he couldn't. "I can't get it over that way," he told Duke. "I'll give you a couple of pictures of it. Can you make a 360 [turn], John?"

"Okay, where do you want? Right here?" offered Young.

"Yeah. Okay, that's fine," said Duke as he fired off a picture. "Could you keep on going around? Let's just make a 360 this way."

"Hey, that's a neat way to take a pan, Charlie," said Young as he realised Duke wanted to use the Rover with its tight 6.1m turning circle as a platform for taking panoramas without having to get off the vehicle. The only thing Duke had to be careful of was to remember to adjust the camera's exposure settings to take account of the varying lighting angle as they spun around.

Interplanetary outside broadcast

NASA had been largely hostile to the idea of TV from the Moon in the months leading up to Apollo 11. But they came to embrace it towards the end of Apollo, endowing the Rover with the ability to send live television from wherever the vehicle happened to stop. Thus the LRV became the most outlandish mobile outside broadcast unit ever. This was at a time when the use of satellites to bounce TV around Earth was an extremely expensive rarity and most broadcasters had to use multiple hops of microwave links to get pictures from their far-flung events. Live TV from the Moon was a remarkably bold move, and technically difficult with the technology of the time. David Scott recalled, "When you think of the number of people involved, you've got to have power to the television [camera], you've got to move it. That may seem simple, but it's not. It's a big deal."

Television brought huge benefits to the Rover missions, not least the enormous PR bonus of "exploration at its greatest" beamed directly into people's living rooms around the world. It also allowed the engineers and scientists at Mission Control to directly participate in the astronauts' work and, thanks to its remote control capability, it let the two guys on the surface of the Moon get on with more important tasks than operating a TV camera. And, as Scott later said, it offered them "a third pair of eyes on the Moon".

Earlier Apollo missions had tended to place the TV camera in one static place, pointing in the general direction they thought the action was going to unfold. This made for pretty boring

ABOVE Charlie Duke's shots that form this panorama show John Young aligning the high gain antenna towards the end of their time on the Moon. Next to him is the TV camera which allowed Mission Control to participate in their exploration from 400,000km away. *(Charlie Duke/NASA/David Woods)*

TV, even though it was live from the Moon! But with the LRV's mobile camera platform the scene changed each time the Rover stopped at a new site. As Scott pointed out excitedly long after his mission, "Each place you go you're going to have a whole new scene, a whole new set of circumstances that you can get people involved in."

But first, they had got to get it working. The camera had to be connected to its control unit and the signals had to go to the LCRU which would send them to Earth only if the dish antenna was properly aimed. As part of loading their newly deployed Rover on their first day out, Scott and Irwin installed the LCRU, the high gain antenna, and the TV camera with its control unit on the very front of the vehicle. "See if I can find the Earth," said Scott as he tried to centre the planet in an aiming sight on the dish for the first time. He found that Earth's image through the sight was much dimmer than expected. "There she is. Okay; pointed right at you, Houston."

"Roger. No TV yet, but we're looking," said Capcom Joe Allen.

Everyone in Mission Control was eager to receive a picture, especially engineer Bill Perry and his supervisor Olin Graham who had designed the control system that would let flight controller Ed Fendell operate the camera from his console. They had already seen the camera's output when it was plugged into the Lunar Module's own communication system, but not yet through the Rover's LCRU. Scott remembered the tension of the moment. "A lot of people are holding their breath to see if the TV will work when it's on the Rover. First time, and for all the guys who will control it and run it, it's a big deal. Big step in lunar exploration."

"Dave, we want you to verify the CTV switch, 'On', please," checked Allen, "and that the high gain's pointed, in TV Remote."

"Okay, Joe; stand by. I'll go look."

Scott rechecked the antenna, the LCRU settings and then the power switch on the camera's control unit.

"Presto chango; there's the TV," cheered Allen as the signal worked through the communications links from the Moon to the large receiving dishes, then across Earth and through the processing that turned it into a standard TV signal.

"Oh, beautiful, I'm glad to hear that," said Scott.

Immediately, Ed Fendell gained control and began to move the camera on its pan-and-tilt mount and control exposure and zoom while the astronauts continued to prepare the Rover by installing a pallet full of tools and collection bags. Never before had people on Earth had the freedom to look around an alien landscape in the way Fendell was doing now. Allen couldn't help himself. "The TV scenery for us is breathtaking."

"Good," said Scott, recognising that Allen's spontaneous comment meant that the experience was in some way being shared. "Can't be half as breathtaking as the real thing

ABOVE The photographs for this panoramic vista were taken from the lower slopes of Mount Hadley Delta by Jim Irwin at Station 2 on Apollo 15's first outing. David Scott is working at LRV-1 whose TV camera, itself a novel feature of this mission, is giving viewers on Earth a view of the spectacular scenery that the Hadley Plain offers. Beyond the Rover, Hadley Rille snakes off into the distance. It is 1,500m wide and 300m deep. To its right is the partly shadowed peak of Mount Hadley rising over 4km above the plain.
(Jim Irwin/NASA/David Woods)

RIGHT A frame from the TV coverage at Apollo 15's Station 2 that Capcom Joe Allen marvelled as being "absolutely unearthly".
(NASA)

though, Joe; I'll tell you. Wish we had time just to stand here and look."

"It gets everybody involved," commented Scott afterwards. "And there's nobody in the control centre who isn't glued to this. I mean, absolutely glued to it! So it brings everybody in. It makes them all part of the deal, which I think is one reason why the bosses probably put the television camera on."

The suit techs could see how their suit had been configured and if any part of it was not quite as it should be. Everyone in Mission Control could see how the crew were performing and whether they were having any issues. But in particular, the geologists, whose trained eyes can read the rocks and the landscape, could look around and explore separately from the astronauts.

At Apollo 15's second geology station, Mission Control was still getting used to the camera. Scott and Irwin were sampling rocks

RIGHT This TV frame was after Ed Fendell had zoomed in on the far-off meanders of Hadley Rille. While the crew get on with their work, Earthbound geologists can see individual boulders and outcrops in the rille.
(NASA)

LEFT TV frame of 'Big Muley' showing the light markings that attracted the geologists' attention. *(NASA)*

just above the point where Hadley Rille, a canyon some 1.5km wide, took a sharp turn to the northwest which gave them a truly spectacular view of the rille as it snaked off into the distance. When Fendell aimed the TV camera along its length, Joe Allen couldn't help himself. "And we have a view of the rille that is absolutely unearthly."

"Yeah. Didn't we tell you?" said Scott. "Tell me this isn't worth doing, boy."

Without interrupting the astronauts, the geologists were able not only to see a beautiful alien vista, but they could immediately glimpse the boulders scattered across the floor of the rille and the outcrops that hinted at the layers beneath the adjacent plain. The view was gold dust to the experts gathered in the science support room at Mission Control.

On Scott and Irwin's second outing, Scott was busy describing a sample he had in his hands when the TV spotted a sample bag fall from Irwin. "And, Jim, you may have dropped your sample bags," warned Allen.

"Yeah. I dropped one," agreed Irwin.

"Joe, thank you," giggled Scott.

"I don't know what we would do without you, Joe," said Irwin.

Almost a year later, at Apollo 16's first geology station, Young and Duke were working beside the rim of a small but deep crater called Plum when the geologists in the science support room became interested in a prominent rock right in front of the TV camera; thanks to a light-toned patch at the top of it. They passed their interest to a dedicated flight controller with the moniker 'Experiments' in the Mission Operations Control Room. He then ran the request past the flight director known as 'Flight' who was in charge of the mission, at that time Pete Frank.

"Flight, Experiments." This was the standard way of addressing someone across the

BELOW On LRV-2's first outing on Apollo 16, John Young took the photographs for this panorama that shows Charlie Duke standing next to Plum Crater. To the left is Flag Crater. The Rover is sitting at the far side of the crater with the rock that would be named 'Big Muley' to its right. *(John Young/NASA/David Woods)*

RIGHT Charlie Duke checks with Mission Control that it is the rock they are interested in. TV frame. *(NASA)*

FAR RIGHT In this TV frame, Duke is rolling 'Big Muley' up his leg. *(NASA)*

RIGHT With 'Big Muley' firmly in his grasp, Duke approaches the Rover to stow it in a space under one of the seats. *(NASA)*

RIGHT Apollo sample number 61016, otherwise known as 'Big Muley', an 11.729kg, 4 billion-year-old piece of rock composed largely of plagioclase that has been shocked during a much more recent impact. *(NASA)*

RIGHT A TV frame from Apollo 15 shows how sunlight coming from the left caught small amounts of dust on the lens and affected the picture, giving it an excessively milky appearance. *(NASA)*

communication loops: say who you are calling and then say who you are.

"Go ahead," said Frank.

"Looks like they'll be going by that rock," pointed out Experiments. "Could we ask them to pick it up on the way back?"

Frank made a call on it and passed on the request to Capcom Tony England, the only point of contact with the crew.

As Duke approached the Rover, England completed the chain, requesting: "As you come around there, there is a rock in the near field on this rim that has some white on the top of it. We'd like you to pick it up as a grab sample."

Duke pointed at it with his scoop. "This one right here?"

"That's it," confirmed England.

"That's a football-size rock," exclaimed Duke as he dropped to his knee to pick it up.

"Are you sure you want a rock that big, Houston?" checked Young.

"Yeah, let's go ahead and get it," said England as Duke manhandled the 11.7kg sample to his knee and then up his thigh.

"If I fall into Plum Crater getting this rock," teased Duke, "Muehlberger has had it."

Bill Muehlberger was the chief geologist on the mission and this rock, the heaviest single sample returned from the Moon, was named 'Big Muley' in his honour.

The TV camera gave few problems, but chief among them was dust on the lens. "Jim," called Allen, on day 2 of Apollo 15's expedition. "Could you dust our lens?"

Mission Control had been discovering that even the slightest coating of dust on the lens could catch the sunlight and wash out the camera's image. "Stand by," said Irwin. "Point up and look at me, and I'll clean you off." Though the astronauts had a lens brush, the effect of the dust was far worse than had been anticipated. But the camera didn't move and, initially at least, Irwin was wary of forcing it by hand.

"Jim, could you give us a little help on the tilt," requested Allen. "We seem to be hung up."

The camera's tilt axis was below the camera itself, which meant that if ever it was tilted away from horizontal, part of the camera's weight became a force to tilt it further. On Apollo 15, the conditions on the Moon showed up deficiencies in this mechanism when the clutch failed to provide sufficient bite to overcome the camera's weight. Tilt it too far, and Mission Control were left looking down at the ground or up in the sky. Irwin manually tilted the camera and proceeded to clean the dust off its lens.

"How does that look to you?" he asked.

"Would you check the oil, too, please?" joked Allen.

At the VIP site

Having an independent TV camera mounted on the Rover gave Houston and everyone else on Earth an extra treat – a ringside seat at each mission's second launch. The first was on Earth at the Kennedy Space Center. The second would be on the Moon as the top half of the Lunar Module lifted off at the start of the astronauts' journey back to Earth.

Since the spacecraft was going to rise only a few tens of metres before tilting over and flying to the west, the plan was to park the Rover 100m to the east from where it would have the best chance of observing the launch. Knowing that the Rover may never move from that spot again, the Apollo 15 crew called this activity 'Rest in Peace'.

On Earth, many dignitaries and press watched the launch of the huge Saturn V rocket from a position 5km from the pad known as the VIP (for Very Important Person) site. After Apollo 15, this term was used for the Rover's final parking spot because it was from there that everyone was going to watch a launch. But the 3sec delay between issuing a command to the camera and seeing the response would make it difficult to track the rising spacecraft.

The faulty tilt clutch on LRV-1's camera mount dashed any hopes of following Apollo 15's

ascent. Fendell didn't dare command the camera to tilt, and instead viewers on Earth had to be content with watching the spacecraft lift-off from the rest of the Lunar Module and pop smartly out of frame. It was a spectacular sight nonetheless.

Fendell got his first chance to track the spacecraft's rise on Apollo 16. Knowing the precise time that lift-off was due, engineer Harley Weyer planned a set of cues for Fendell that would take account of the delay and tell him when to press the buttons in Houston. In theory, these would synchronise the camera's tilt to the LM's climb to orbit. But Weyer's calculations assumed that the Rover's resting place was the correct distance from the LM and also level. In Apollo 16's case it was not accurately positioned and the rough, sloping ground spoilt efforts to track the LM properly.

"Actually, with Apollo 16, it was John Young's fault," Fendell remembered. "He parked the Rover in the wrong place. For the last mission, I told Gene Cernan, 'Make sure you line that Rover up in the right place so I can film the lift-off!'"

Despite this, Fendell got a good view of

ABOVE The Mission Operations Control Room (MOCR) at the moment of Apollo 15's lift-off from the Moon, 2 August 1971. At the Capcom console in the striped shirt is Apollo 14 astronaut Ed Mitchell. *(NASA)*

LEFT Flight controller Granvil Pennington watches Apollo 15's lift-off, unable to track it due to the camera's faulty clutch. *(NASA)*

ABOVE LRV-3 at the VIP site, ready to cover their lift-off later that day. The rear wheels are exposed as he has removed their fender extensions, including his replacement fender. *(Gene Cernan/NASA)*

the first few metres of the ascent; sufficient to see that the shock of the engine ignition had disrupted the thermal blankets on the back of the ascent stage, and Mission Control was able to ask Ken Mattingly, the astronaut who had remained in orbit, to take a good look at them when his companions got closer.

Fendell's triumph came on Apollo 17 when all the variables lined up: the Rover was level and the timings matched well enough that Fendell was able to track the ascent stage of *Challenger* upwards and right through 'pitchover', the moment when the LM tilted over and began to travel horizontally. "On my last chance I managed to get it right! A perfect lift-off from the lunar surface caught perfectly by the TV camera!"

The camera panned up as far as it could to watch *Challenger* heading off into the black sky, and then tilted back down to the lunar dust – where it found one of the last footprints ever placed on the Moon. From then on, the only role left for the Rover and its camera was to watch the landscape through the remainder of its fortnight-long daytime as long as the system could hold out in the increasing heat of

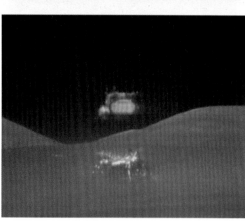
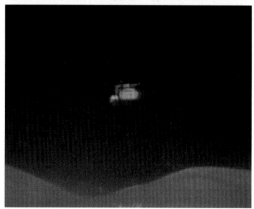

RIGHT This sequence of TV frames shows the ascent stage of *Challenger* at lift-off and through its first moments of ascent as Ed Fendell tracks it. The colours imparted to the flying debris reveal the scanning sequence of the TV camera: red, blue then green. *(NASA)*

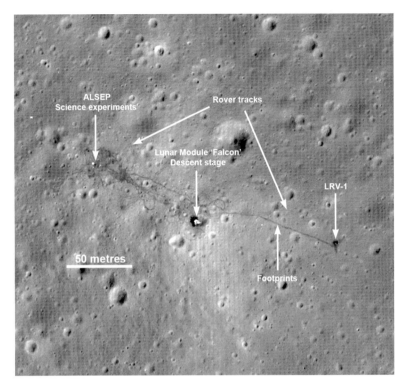

ABOVE The Apollo 15 landing site, taken on 6 November 2011 by Lunar Reconnaissance Orbiter. Even after many decades, the tracks left by the astronauts and their Rover are plain. The final resting position of LRV-1 to the east is clearly seen. *(NASA/GSFC/Arizona State University)*

ABOVE Apollo 16's landing site, taken on 6 November 2011 by Lunar Reconnaissance Orbiter with the final parking position of LRV-2. *(NASA/GSFC/Arizona State University)*

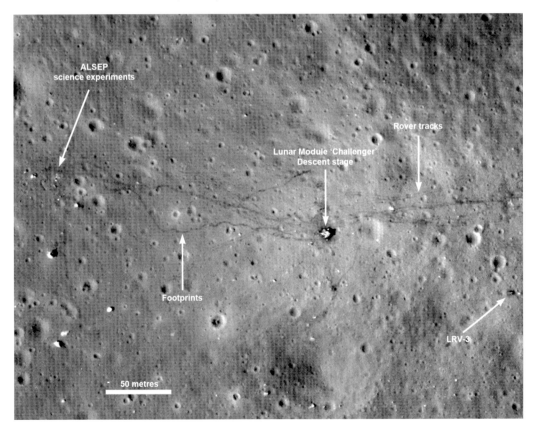

LEFT The Apollo 17 landing site, taken on 14 August 2011 by Lunar Reconnaissance Orbiter showing LRV-3 in its final position. *(NASA/GSFC/Arizona State University)*

ABOVE After he had parked LRV-1 for the final time and prepared its battery covers, David Scott took a series of photographs for this 'Rest in Peace' panorama of the Rover with Mount Hadley beyond. *(David Scott/ NASA/David Woods)*

RIGHT An enlargement of the 'Rest in Peace' panorama shows the Bible that Scott left on the T-handle. *(NASA)*

lunar noon. All the Rovers succumbed to the conditions after a few days, with Apollo 16's lasting the longest at six days.

Rest in peace

For more than 40 years the three operational Rovers have sat immobile where they were left by the departing astronauts. There is no hint that anyone is about to revisit them soon. After Apollo, flights to the Moon seemed too expensive for a world whose attention was moving elsewhere, and it wasn't until 2009 that we caught another glimpse of these venerable vehicles.

The early 21st century saw a renaissance in lunar interest, when various countries sent automated probes to study it from orbit. Notable among them is NASA's Lunar Reconnaissance Orbiter, which carries powerful cameras to map the entire surface. Its largest camera can resolve objects less than 50cm in size, and it wasn't long before the Apollo sites were targeted by the mission planners. For the first time, the exact paths taken by the astronauts in their Rovers could be directly traced across the lunar terrain, rather than being guessed at from other evidence. The disturbance of the soil by the astronauts' boots and the distinctive double lines left by the Rovers' wheels are still clearly visible, and they will probably persist for many thousands of years until degraded by the incessant rain of micrometeoroids.

The –200°C chill of hundreds of lunar nights and the blasting heat of each lunar noon has probably long ago wrecked the LRV batteries. The plastics used in the thermal blankets may have perished by now, but otherwise, the Rovers are very much as the crews left them.

David Scott placed a Bible on the T-handle of LRV-1 and Gene Cernan believes he may have left one of the Hasselblad cameras behind on LRV-3, intentionally pointed to the sky to record micrometeorite impacts on the surface of its lens; a future project for some researcher who maybe isn't born yet.

As happened on all the Rovers, Young and Duke gave LRV-2's radiator panels a good dusting before they left. But Houston had an additional chore to leave the little car tidy for its next owner.

"Charlie, we'd like you to dust the top of the console," instructed Tony England, referring to the white control and display console.

Young queried the instruction. "Why do you want to do that, Houston?"

"We want to keep the temperature of the panel down."

"In case anybody comes back?" said Young.

"I guess so," said England. "Keep it nice for the next guy."

With no current plans to send the next guy, the question of whether there will be anything useful remaining when they eventually get there is a moot point. Probably by the time anyone returns, the Rovers will be more important as relics of the first episode of human exploration of another world.

As this final chapter in the history of the Apollo Lunar Roving Vehicles proves, these lovingly crafted cars were triumphant in fulfilling their missions; performing well beyond expectations. The astronauts of Apollo 17 summed up this success when concluding their post-mission report. "The Rover is an outstanding device which increased the capability of the crew to explore the Taurus-Littrow region and enhanced the lunar surface data return by an order of magnitude and maybe more."

They were born in an era of unparalleled engineering creativity that took humans to another world for the first time. And they were the epitome of that creativity; extremely well suited for their role. Not wasteful but clever, efficient and expressing the compromises that every engineer has to make when faced with the stresses of time, money, and expectation. Like the extraordinary Lunar Module that took them to the Moon, their beauty lay not in their form but in the role they performed, and in the intelligence, perspiration and humanity of those who designed and built them, and those who drove them on the Moon.

"One of the finest running little machines I've ever had the pleasure to drive."
Gene Cernan, 14 December 1972

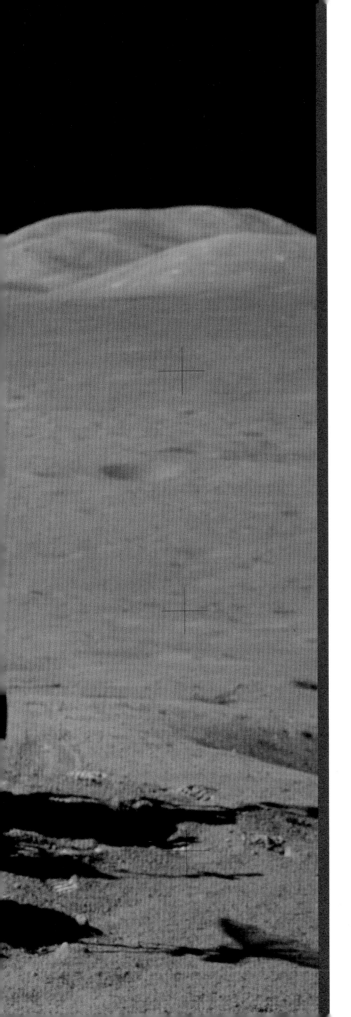

Chapter Eight

Epilogue

In 2003 engineers embarked on a project to recreate 12 replica wire-mesh wheels from the Lunar Roving Vehicle. Their objective was to better understand Bekker and Pavlics' ingenious airless, all-terrain tyre technology, for further applications both on and off Earth.

OPPOSITE A portrait of the LRV-3 parked at the Taurus-Littrow landing site, showing the TV camera pointed off to the left and the high gain antenna pointed towards Earth, which is over the South Massif. *(Gene Cernan/NASA)*

LEFT The placard mounted on the Control and Display Console of LRV-1 reads "Man's First Wheels on the Moon, Delivered by Falcon, July 30, 1971." The signatures of the three members of the Apollo 15 crew are at the bottom. In the background we can see Dave Scott.
(Jim Irwin/NASA)

Perhaps surprisingly, given the advancements in material science and the extra engineering knowledge available to them, the team found the LRV's wheels difficult to replicate. Their efforts served to emphasise just how hard a task it must have been back in 1969 to create the car that carried a total of six astronauts a distance of almost 100km across the Moon.

Those who rose to this challenge all those decades before, paid a high price to design and manufacture the LRV to such a tight schedule. And many can still remember the toll it took on their health, their family lives, and their marriages. "I can honestly say there weren't many light moments in the programme," reflects Sonny Morea, NASA's Rover project manager. "Every day was highly stressed."

At the same time however, they also recall those hectic, taxing years as some of the most gratifying of their lives. Ferenc Pavlics spoke for many when he summed up his experiences on the programme for aerospace historian Anthony Young. "The most rewarding part for me," he clearly remembered, "was that the teamwork was so excellent. Everyone was pulling for the project, for its success. I didn't hear a single complaint about overtime or working weekends. Everyone was enthused and excited about the

LEFT This self-portrait of NASA's Mars Exploration Rover *Opportunity* was captured by the Rover's front hazard avoidance camera. The dramatic snapshot of the Rover's shadow was taken as the vehicle advanced into Endurance Crater. The image was taken on sol 180 (26 July 2004), marking a doubling of the hardy Rover's primary 90-sol mission. As of August 2012 *Opportunity* was still functioning on Mars, over eight years after landing on the Red Planet.
(JPL/NASA)

programme. I think that made it possible to complete it in such a short time."

For many of them the excitement of driving a vehicle across another planet was the crowning glory of Apollo, and one they were proud to have participated in. With its live television feed, the Rover had enabled unprecedented coverage of these unique expeditions to the mountains of the Moon; once again creating a sense of wonder at what human beings can achieve.

Without the three LRVs, none of the major scientific discoveries of Apollo 15, 16 and 17 would have been made, and as Harrison Schmitt, the only scientist to set foot on the Moon, puts it, "our current understanding of lunar evolution would not have been possible." More than this, the Rover enabled samples to be obtained that underpin our understanding of the origin of Earth and the Solar System.

This scientific legacy will long be remembered in academic journals, but the wider contribution of the Lunar Roving Vehicle to human history is often overlooked. In developing them, the engineers at Marshall gained a unique insight into designing hardware for human spaceflight. Although the LRV was not as sophisticated as a habitable laboratory, it functioned as a small, extra-terrestrial workstation and had similar crew requirements. The knowledge and experience gained was applied to the design of crew systems for later prestige projects like Skylab, the Space Shuttle, and the International Space Station.

Its drive-by-wire controls informed the design of future Earth-based vehicles and the multi-purpose 'joystick' was later adapted for computer games. A variation of it was also developed to permit paraplegics to drive without making extensive modifications to a vehicle.

The Lunar Rover's engineering legacy has reached far beyond the Earth too, into the systems of a new generation of robotic exploration rovers designed for more distant planetary surfaces. Since 1997 NASA has successfully dispatched four six-wheeled rovers to the red planet, most recently with their giant Mars Science Lab *Curiosity* in 2012. Their cumulative mileages will soon exceed that of the three LRVs on the Moon From their dead-reckoning and semi-autonomous navigation capabilities, to their hardy metal wheels and their stereo high-definition cameras, these planetary rovers have all borrowed from the original technological innovations pioneered for the LRV. Indeed, Ferenc Pavlics and Ron Creel, two of the original Rover designers, continue to play an active part in the design of this new breed of interplanetary explorers.

This is where the real legacy of the Lunar Roving Vehicle lies; as a potent symbol of humanity's eternal quest to cross that next horizon. The Apollo LRVs epitomised our yearning to explore. And perhaps no one expressed this sentiment better than the man who drove the first one on the Moon, as he stepped off the footpad of Lunar Module *Falcon* onto the lunar surface for the first time.

"As I stand out here in the wonders of the unknown at Hadley, I sort of realise there's a fundamental truth to our nature. Man *must* explore. And *this* is exploration at its greatest."

BELOW The first self-portrait of *Curiosity* taken on August 8 2012. The back of the rover can be seen at top left, with two of the rover's right side wheels on the left. Part of the pointy rim of Gale Crater forms the lighter color strip in the background. Bits of gravel are visible on the deck of the rover, blasted there by the rocket motors from the hovering sky crane which lowered it to the surface hours before. *(NASA)*

Glossary

ABMA — *Army Ballistic Missile Agency.* Rocketry development agency headed by Wernher von Braun.

AC — *Alternating Current.* Term for the flow of electricity where the direction of flow keeps reversing.

Ackermann — A type of steering which allows the inside front wheel to turn through a greater angle than the outside wheel when cornering, thus avoiding the tyres slipping sideways when following a path around a curve.

ALSEP — *Apollo Lunar Surface Experiment Package.* Scientific instruments deployed on the lunar surface by Apollos 12 and 14–17.

AOT — *Alignment Optical Telescope.* An optical instrument in the LM. It had a guard rail to which lamps were clamped. These clamps were used to repair the fender of LRV-3.

BSLSS — *Buddy Secondary Life Support System.* This was a hose arrangement that allowed a crew on the lunar surface to share cooling water in an emergency. It was kept in a bag on the back of the Rover's seats.

Bus — In electrical engineering, a bus refers to a common conductor that feeds power or signals between a number of sources or destinations.

CM — *Command Module.* The part of the Apollo spacecraft where the crew was housed for most of the mission.

CMP — *Command Module Pilot.* Third crewman on Apollo missions, who remained in lunar orbit aboard the CM during the LM's excursion to the Moon's surface.

CSM — *Command and Service Module.* The unitary Apollo spacecraft right up until the time of re-entry.

DC — *Direct Current.* Term for the flow of electricity where the direction is one-way only.

DCE — *Drive Control Electronics.* Electronic package that controlled power delivered to the Rover's wheels based on the commands from the T-handle.

DGU — *Directional Gyroscope Unit.* The gyro-based device that measured the Rover's orientation with respect to north.

Ejecta — The material that is ejected from a site during the formation of a crater. Such is the energy of a cosmic impact that it forms an explosion. The impactor is generally vaporised while the target rock is blasted outwards as ejecta.

EVA — *Extravehicular Activity.* The practice of leaving the pressurised confines of a spacecraft to go outside wearing a spacesuit.

GCTA — *Ground Commanded Television Assembly.* Overall name for the TV camera, its mount and its control box, the TCU.

GM — *General Motors.* Major subcontractor to Boeing for the LRV.

GSFC — *Goddard Space Flight Center.* NASA centre responsible for the operation of the Lunar Reconnaissance Orbiter.

HTC — *Hand Tool Carrier.* A small, three-legged truss structure that held tools used on the surface, such as hammer, corer, shovel and tongs. It also had a bag in the centre for rock samples.

INCO — *Instrumentation and Communications Officer.* Mission Control position responsible for the operation of the TV camera.

IPI — *Integrated Position Indicator.* A circular display that indicated the LRV's heading. It included numeric displays for range, bearing and distance travelled.

King pin — The pivot that actuates the steering of a wheel.

KSC — *Kennedy Space Center.* Named following the assassination of President Kennedy, this refers to the facilities that were laid out at the north end of Merritt Island in Florida to support Apollo/Saturn launches. It lies next to the military rocket launch site at Cape Canaveral, itself temporarily renamed Cape Kennedy.

LCRU — *Lunar Communications Relay Unit.* An electronic package mounted on the Lunar Rover that dealt with all communication between the crew and mission control. Functions included voice, biomedical, telemetry, television and remote control commands for the TV camera.

LEC — *Lunar Equipment Conveyor.* A looped strap affair with a simple pulley used on the first two landings to transport equipment and samples between the LM cabin and the lunar surface.

LM — *Lunar Module.* The two-stage landing craft used to take crewmen from lunar orbit to the surface, sustain them during their exploration and return them to the CSM. Pronounced as 'lem'.

LMP — *Lunar Module Pilot.* Second crewman in the lunar module. He did not actually pilot the LM, but

	acted as a flight engineer and co-pilot, aiding the commander in the LM's operation.
LRO	*Lunar Reconnaissance Orbiter.* Spacecraft launched in 2009 to map the Moon in detail. Among its tasks, it has imaged all the Apollo landing sites and shown the tracks of the astronauts and the Rovers.
LRV	*Lunar Roving Vehicle*. Electric-powered car that was added to the final three lunar landing missions to facilitate a greater range of surface exploration.
Mare	The scientific name derived from Latin given to the smooth dark areas of the Moon, commonly called 'seas'. Pronounced as 'maa-ray'.
Maria	The plural of mare.
MESA	*Modularized Equipment Stowage Assembly.* Unit built into the Lunar Module to carry equipment and tools required for surface exploration.
MET	*Modular Equipment Transporter*. A two-wheeled cart used on Apollo 14 to carry tools, cameras and rock samples during the astronauts' walking traverses.
MOCR	*Mission Operations Control Room.* Pronounced to rhyme with 'poker'. Often termed 'mission control', it was the hub of the flight control effort during a mission and was supported in the task by various nearby rooms.
MOLAB	*Mobile Laboratory.* Early, but seminal, NASA design for a pressurised lunar rover.
MSFC	*Marshall Space Flight Center.* Usually just called 'Marshall' and located in Huntsville, Alabama, this NASA centre was primarily concerned with the development of launch vehicles. It was given the job to develop the Lunar Rover.
MTA	*Mobility Test Article.* Vehicles designed by Bendix and Boeing for NASA in 1964.
NASA	*National Aeronautics and Space Administration.* American agency that was created in 1958 to maintain US space effort in the civilian realm. In 1961, it was tasked with running the Apollo programme.
NTSC	*National Television Standards Committee.* US body that defined TV standards for the USA in the 20th century.
OPS	*Oxygen Purge System.* An emergency package of oxygen contained in two high-pressure bottles. It was carried on top of the backpacks during lunar surface forays. The CMP also carried it while making a space walk from the CM to retrieve the film canisters during the crew's return to Earth.
PLSS	*Portable Life Support System.* An astronaut's backpack while on the lunar surface.
PWM	*Pulse Width Modulation.* Electronic technique used in the Rover for efficient control of the motors by switching them on and off very fast.
RCA	*Radio Corporation of America.* Company that built the Rover's communication equipment including the LCRU and GCTA.
Regolith	The surface layer of rubble and dust that is draped across the Moon's entire surface. It builds up over huge expanses of time to depths of tens of metres.
RTG	*Radioisotope Thermoelectric Generator*. Power supply for the ALSEP suite of science instruments deployed on five of the landing missions. It generated electricity by direct conversion from heat provided by a radioactive plutonium source via an array of thermocouples.
S-band	A band of radio frequencies from 2 to 4GHz that included those used by Apollo to communicate with Earth from the Moon.
SM	*Service Module.* A cylindrical section of the Apollo spacecraft that housed most of the consumables and propulsion systems.
Spline	A cylinder, perhaps hollow, with ridges that run parallel to its axis. On the Rover, splines were used on the suspension and in the harmonic gear.
SPU	*Signal Processing Unit.* Part of the Rover's navigation system. Like a simple computer, it performed the calculations to keep track of the Rover's whereabouts.
SSE	*Space Support Equipment.* The mechanism built onto the Lunar Module that held the LRV in place and allowed it to be deployed onto the lunar surface.
TCU	*Television Control Unit.* A box beneath the TV camera that controlled the camera's mount and mediated between it and the LCRU.
TEI	*Trans-Earth Injection.* A major rocket burn made to send a spacecraft on a trajectory to Earth.
TLI	*Translunar Injection.* A major rocket burn made to send a spacecraft on a trajectory from Earth orbit to the Moon.
UHT	*Universal Hand Tool*. Tool carried on the lunar surface that was designed for a variety of tasks.
USGS	*United States Geological Survey.* The US agency whose Astrogeology branch was closely involved with the geology carried out from the Rover.
VHF	*Very High Frequency.* A band of radio frequencies from 30 MHz to 300 MHz that includes those used by the crew to speak to each other on the lunar surface.
Zero-phase	A point for an observer that is directly opposite the Sun. On the Moon, there was a particularly bright reflection near zero-phase that made driving down-Sun difficult.

Acknowledgements

The idea for a Haynes Manual on the Lunar Roving Vehicle came to Philip Dolling and Chris Riley in 2009, when they were writing the *Apollo 11 Owners' Workshop Manual*. But until their first Workshop Manual on a spaceflight topic had proved itself, the idea of an LRV book was put on the back-burner. Over two years later, towards the end of 2011, it was David Woods, editor of NASA's *Apollo Flight Journal* and author of *How Apollo Flew to the Moon*, who approached them with the idea again, and together the three authors began discussions with Haynes about a *Lunar Rover Owners' Workshop Manual*.

In early 2012, the project received the green light, and work began. Once again Philip Dolling brought his deep experience of story telling to the project, carving out time in a hectic schedule to read the drafts.

The pressure for us to deliver this book within six short months paralleled, in a much smaller way, the story of the Rover itself. And, again like the Rover, it was only thanks to many years of scholarly work and expertise of those who'd trodden this path before, that such a tight deadline was possible to achieve. With the flexibility and support of the team at Haynes, and the expert eye of our copy editor David Harland we were just able to deliver the book in time.

We are also particularly grateful to Ron Creel, Ferenc Pavlics and Otha 'Skeet' Vaughan, from the original Apollo LRV engineering team, who all gave generously of their time to read and comment on our manuscript. Their scrutiny of our story has ensured that it is as accurate as possible. The scholarly work of Apollo historian Anthony Young also proved vital in getting the story straight, and we are grateful to him for allowing us to quote from his own book *Lunar and Planetary Rovers*. Accounts from the Apollo historian Eric Jones, creator and editor of the remarkable *Apollo Lunar Surface Journal*, also appear in this book, and we are most grateful to him for his inspiration and unquenchable determination to document this unique period in human history.

At NASA's Marshall Historical Reference Collection, Molly Porter spent considerable time tracing obscure documents and pictures for us, and at NASA's Johnson Space Center, Mike Gentry, Connie Moor, Jody Russell and Jim Brazda also made time in their extremely busy days to help us trace some of the images appearing in the book. Gwen Pitman at NASA HQ in Washington and Vivake Asani at NASA Glenn in Cleveland were also supportive of this effort, as was Mary Ann Hager, at the Lunar and Planetary Institute in Houston, Texas. At the University of Alabama in Huntsville, Anne Coleman arranged access to high-resolution versions of Rover technical documents.

At the British Interplanetary Society in London, the authors would like to thank Mark Stewart, Bob Parkinson, Suszann Parry and Mary Todd, for support and access to their own remarkable archives. Long-time friend and Haynes author David Baker, and his own friend and fellow author Rob Godwin also generously shared their knowledge and experience in helping us to source some of the illustrations and photographs which appear here. Other photographs and drawings were kindly supplied by fellow Apollo enthusiasts Karl Dodenhoff, Mike Jetzer and Scott Schneeweis.

We are also very grateful for the patronage of Apollo 15 Commander David R. Scott, who has been a supportive friend of our many Apollo-related projects over the years, and whose generous endorsement of our book in his foreword makes us proud, thrilled and happy!

As always with such projects, the support of an author's family is an essential element for the completion of any book. Chris would like to thank his mother Stephanie Riley, for diligently proof reading and further improving our manuscript, and his wife Jacqui and daughter Evelyn for their patience and understanding. David would like to thank his very long-suffering, yet endlessly supportive Apollo-geek's wife Anne and his two sons Stephen and Kevin, and Phil Dolling would like to thank his wife Sara and his children Beatrice and James. We can only hope that they and others of their generation will one day visit the Moon for themselves and maybe go for a drive.